S0-DJF-968

Military Livings™

Military Space-A Air Basic Training™ and Reader Trip Reports

L. Ann Crawford
Vice-President, Military Marketing Services, Inc.
and Publisher, Military Living Publications

William Roy Crawford, Sr., Ph.D.
President, Military Marketing Services, Inc.
and Military Living Publications

VICE-PRESIDENT - MARKETING - R.J. Crawford

EDITORIAL ASSISTANTS - Leon G. Russ and Bryan Suit
ELECTRONIC ARTIST - Jeremy Rodgers
COVER DESIGN - Lindsey Tozier-School

CHIEF OF STAFF - Nigel Fellers
OFFICE STAFF- Eula Mae Brownlee, Anna Belle Causey, Irene Kearney, Rose McLain, Lourdes Medina, Joel Thomas, MSG USA (Ret), Modest Zacharczenko.

Military Living Publications
P.O. Box 2347
Falls Church, Virginia 22042-0347
Tel (703) 237-0203
FAX (703) 237-2233

NOTICE

The information in this book has been compiled and edited either from the activity/facility listed, its superior headquarters, or from other sources that may or may not be noted. All listed facilities, their locations, hours of operation, and telephone numbers could change. Flight schedules, including destinations, routings, frequency of flights, and aircraft used, are always subject to change. Space-A passenger eligibility could change; however, we have published the most-up-to-date information available. The "how to travel Space-A" supporting information in the appendices is subject to change, but the latest changes to these appendices were included at press time. This book should be used as a guide to Space-A travel with all of the above in mind. **Please forward any additions or corrections to the publisher.**

This guide is published by **Military Living Publications,** a private firm in no way connected with the U.S. Federal or other governments. The guide is copyrighted by L.Ann and William Roy Crawford, Sr. Opinions expressed by the publisher and writers herein are not to be considered an official expression by any government agency or official.

The information and statements contained in this book have been compiled from sources believed to be reliable and to represent the best current opinion on the subject. No warranty, guarantee, or representation is made by **Military Living Publications,** as to the absolute correctness, or sufficiency of any representation contained in this or other publications and we can assume no responsibility.

Copyright 1994

L. Ann and William Roy Crawford Sr.

First Printing - September 1994

MILITARY LIVING PUBLICATIONS
MILITARY MARKETING SERVICES, INC.

All rights reserved under International and Pan-American copyright conventions. No part of this book may be reproduced in any form without permission in writing from the publisher, except by a reviewer who wishes to quote briefly from listings in connection with a review written for inclusion in a magazine or newspaper, with source credit to **Military Living's Military Space-A Air Basic Training™ and Reader Trip Reports**. A copy of the review when published should be sent to Military Living Publications, P.O. Box 2347, Falls Church, VA 22042-0347.

Library of Congress Cataloging-in-Publication-Data

Crawford, Ann Caddell.
Military Space-A air basic training and reader trip reports / Ann Crawford, William Roy Crawford, Sr.
p. cm.
ISBN 0-914862-38-3
1. United States--Armed Forces--Transportation. 2. Air Travel. 3. Military dependents--Transportation. I. Crawford, William Roy, 1932- . II. Title.
UC333.C7254 1994
355.3'4--dc20 94-23130
CIP

ISBN 0-914862-38-3

CONTENTS

SECTION I: AN IMAGINARY OVERSEAS SPACE - A AIR TRIP WITH THE FAMILY

SECTION II: AN IMAGINARY SPACE-A AIR TRIP BY CONUS MEDICAL EVACUATION (MEDEVAC) FLIGHT

SECTION III: AN IMAGINARY SPACE-A AIR TRIP BY GUARD AND RESERVE PERSONNEL

SECTION IV: SPACE-A ELIGIBILITY AND DOCUMENTATION REQUIREMENTS TABLES

SECTION V: READER TRIP REPORTS

APPENDICES

DEDICATION

Military Space-A Air Basic Training is dedicated in memory of Lieutenant Colonel Robert H. (Bob) Scoggin, U.S. Army Retired. Most cartoons in this book were done by Bob as he sat in miltary air terminals around the world waiting for his next flight.

Front & Back Cover Photo Credits

A C-17 takes off - Photo by U.S. Air Force
John & Ursula Wooten enjoy the ride to Iceland on A C-135 - Photo courtesy of John Wooten
Space-A Passenger Service Personnel, Tsgt Peeples, Mr. Goodman, & SrA Carpenter at Hickam, Hawaii - Photo by Ann Crawford
A C-5B at Dover AFB, Delaware - Photo by Ann Crawford
Les Swanson waits for his flight at Mildenhall, England - Photo courtesy of Les Swanson
Space-A passengers board a C-5 Galaxy at Tinker AFB, OK - Photo by Larry Lang, Soldiers Magazine

INTRODUCTION

If you have never flown Space-A or if you have not flown Space-A recently, ***Military Space-A Air Basic Training™ and Reader Trip Reports*** has important information to help make your next trip successful. The new Space-A rules are entirely different from previous years.

The privilege of flying Space-A on U.S. military aircraft has saved literally millions of military and their family members a great deal of money. More importantly, however, is that this unique privilege has enabled military and their families to take morale-boosting trips that they could not have afforded otherwise.

Have you and your family joined in on the fun? ***Military Space-A Air Basic Training™ and Reader Trip Reports*** will give you the information you need to get started ***"traveling on less per day....the military way!"***

In surveys, military ID card holders have responded that Space-A air travel is one of the privileges that they value most. Rightly so, because being able to fly Space-A gives the military member and his/her family an immediate cash-saving benefit in addition to the important unmeasurable value of being a part of the military family.

Space-A air travel is also important to retirees giving them a continued sense of camaraderie plus the fact that their service to their country is still recognized in a tangible manner.

HOW SPACE-A AIR TRAVEL HAS CHANGED IN RECENT YEARS

The Space-A travel fee has been eliminated thanks to General Colin Powell and General Ronald R. Fogleman, CINC, AMC, and his staff. General Fogleman has also done everything possible to help uniformed service personnel have a simpler and easier Space-A system. His staff has diligently worked on new initiatives which are now a huge success.

Space-A regulations have changed dramatically in the last few years. For the better, too, as many of the bureaucratic procedures have been eliminated. Here are just a few of the changes which you will learn about in ***Military Space-A Air Basic Training™ and Reader Trip Reports.***

In the early years of Space-A travel, potential passengers had to show up at every flight call or be removed from the list. Later, passengers did not have to show up for every flight, but they had to revalidate their records every fifteen days (in person) to indicate they still wanted to fly Space-A.

Sponsors from one place in the country had to make their way to a military air passenger terminal, sometimes across the entire USA, to sign up for a flight from a given location. Sometimes, the passenger would go back to his/her home base and wait while their names made their way up the list and then travel back to the air terminal again hoping they would get a flight. Often, the cost involved in this procedure obliterated any savings on the actual flight. All of that has now changed so that a greatly improved system of sign-up is now in place.

The Military Airlift Command (MAC) and the Strategic Air Command (SAC) have merged into a highly efficient Air Mobility Command (AMC). Especially good news for Space-A passengers is the fact that the AMC now has a wide variety of aircraft, to include tankers. While the former MAC was very user-friendly to Space-A passengers, SAC was not as well organized to serve them. Now, AMC has applied Space-A regulations on a uniform basis. Many tanker locations have changed to other Air Force bases bringing Space-A air travel to new places not previously served by such aircraft. Tankers remain a favorite with Space-A passengers because of their being able to observe actual refueling of other aircraft. There's a special and unique excitement to tanker missions.

Attitudes have improved toward Space-A passengers. While many terminal staff members and aircraft crews have been helpful in the past, it appears that there is a new emphasis on courtesy and helpfulness from the very "top" to the "bottom."

Those wishing to fly Space-A no longer have to appear at the departure terminal to sign up; they can now do this by FAX or letter. The new FAX numbers are included in this book. This remote sign up capability "levels the Space-A playing field" so that Space-A passengers are treated more fairly in having an opportunity to fly.

For those who still prefer to sign up for a military Space-A trip in person, a self-service sign up procedure which is easy to access is still available.

A one-time sign up procedure allows Space-A passengers to keep their original date of being added to the Space-A passenger roster at the originating terminal. If one mission terminates, the Space-A passenger may sign up for a new flight using the original date. Details on this one-time sign up are in this book.

Luggage allowances have been increased to 140 pounds per passenger with two bags not weighing more than 70 pounds each. Even so, remember there are no porters in the middle of an airfield where passengers often debark! Smart Space-A travelers travel as light as possible! Golf clubs or other sporting gear can travel in lieu of one suitcase.

Another reason to travel light is that by doing so, if an opportunity arises for you to get Space-A on an "executive jet" aircraft, the luggage allowed is only 30 pounds.

On AMC flights which originate overseas, Space-A passengers who are eligible family members may now fly with their sponsor to the final destination of a military flight within the United States; they no longer have to deplane at the first stop. Conversely, family members may take flights with their sponsor which make an interim stop within the United States on a mission which continues overseas.

In-flight meals have been changed to a healthy heart variety and breakfast menus have been added. The meals are very reasonable in cost, and we've heard many good comments about them from Space-A passengers.

The wearing of the military uniform by active duty and reserve personnel when traveling on a Space-A basis has been eliminated. At our press time, only the U.S. Marine Corps still had a requirement that their members wear the uniform when flying Space-A. Marines may, however, travel in civilian clothing on Cat B flights (civilian airline contract flights).

Eligible passengers who are handicapped will now find AMC to be a lot more helpful than in previous years.

Passengers will find many improved terminals due to extensive ongoing renovation projects. Many terminals are now brighter and much more comfortable.

No smoking rules now apply to all DoD aircraft missions.

Now that AMC has made all of these improvements, those who have a desire to fly Space-A should learn the rules and customs of flying Space-A in order to be knowledgeable about how this wonderful system operates. This book, ***Military Space-A Air Basic Training™ and Reader Trip Reports***, will give you the information you need to have successful Space-A air travel flights on U.S. military aircraft.

Once you have familiarized yourself with the basic information, you may want to delve into our other publications to truly become a "*Space-A Expert.*" Military Living currently has three other publications to help you do just that. Our all ranks military travel newsletter, Military Living's ***R&R Space-A Report®*** will keep you informed on the latest rules, improvements, locations of new opportunities and more. The reader trip reports are invaluable and form the nucleus of a powerful reader clearing house of information.

Military Living also published an advanced book on Space-A air travel which gives locations of Space-A opportunities and schedules. It is ***Military Space-A Air Opportunities Around The World™***. In addition, Military Living will have a new version of its popular map, ***Military Space-A Air Opportunities Around The World Air Route Map™***, available by early 1995. Check at your military Exchange to purchase these helpful publications. Not only will you save money by doing so, Exchange profits are recycled back into the military community for their recreation programs. If not available, you may, of course, order them direct from Military Living. Phone orders are accepted with Visa, MasterCard or American Express. Please call **(703) 237-0203** for info. Order forms also appear in the back of the book.

On behalf of military families all over the world, Military Living would like to thank the Air Mobility Command for all they have done to improve the life of military ID card holders. Thanks Guys and Gals for letting uniformed services personnel and their families tag along on your missions.

You'll never really know how many hearts have been healed by a visit home, how many elderly parents and other family members have given thanks that their loved one could come home for a visit by Space-A, especially emergency leave travel; how many marriages may have been saved when military members serving apart have had reunions by AMC Space-A flights; or even how military families have become better educated by being able to travel to far-off lands and view countries first-hand. Families flying together form unique bonds and a favorable impression of military life.

AMC's service to those who have also served is a part of military life which makes the bad days be evened out by the good days. **We know how important AMC personnel are - because our readers have told us.** AMC would like to hear from you direct. You may pick up a special form # 253, Air Passenger Comments, at any AMC air passenger terminal or write direct to: **HQ AMC DOJP, 402 Scott Drive, Unit 3A1, Scott AFB, IL 62225-5302.** Thanks again!

Ann, Roy & RJ Crawford

BACKGROUND

The air cargo and passenger capabilities of all the Military Service Departments were greatly expanded during WW II. They have been maintained since that time to support our worldwide peacekeeping military forces, diplomatic missions, unilateral, multinational, and international obligations and commitments. A uniform policy for the administration of this much sought-after air travel benefit was established by the newly created Department of Defense (DoD) in 1947, and communicated through Department of Defense Directive (DoDD): Chapter 4, DoDD 4515.13-R, Space Available Passenger Regulation and its predecessor regulations.

This directive ensures that Space-A air travel priorities and procedures are equitably administered for all of the seven Uniformed Services (U.S. Army, U.S. Navy, U.S. Marine Corps, U.S. Coast Guard, U.S. Public Health Service, U.S. Air Force and the National Oceanographic and Atmospheric Administration).

" I know it's a privilege , but I like to think that I'm providing jobs for more than fifty thousand Americans."

WHAT IS SPACE-A AIR TRAVEL?

Space-A: The term Space-A is used in the military community to mean many things. Space-A is used to describe the use of, or access to facilities and transportation of the military services, after all known required and authorized use and access has been satisfied. The term Space-A is primarily used to describe the availability of air passenger travel to Uniformed Services members (Active and Retired), their dependents and eligible DoD and other civilian employees (when they are stationed overseas). **(See Eligible Passengers, DoDD 4515.13-R, Paragraphs 4-4, 4-5 and 4-6 and Section IV: Space-A Eligibility and Documentation Requirements Tables.)**

Space-A is also used to describe travel on military trains (discontinued in Germany in 1991), buses, temporary military lodging, or use of military recreational facilities which are available after all required and authorized uses have been satisfied.

The DoD has described Space-A air travel as a by-product of the DoD's primary mission, which is the movement of space-required military cargo and passengers. This means that space not required for the movement of official cargo and passengers can be used for the travel of Uniformed Services members around the world, and their dependents on flights with overseas destinations.

SPACE-A AIR TRAVEL STUDY AND TRAINING

SPACE-A AIR BASIC TRAINING: All uniformed service members undergo elementary or basic training. If you hope to save thousands of dollars on air fares, you too will need some basic training or at least, a refresher course, because policies, procedures and techniques of Space-A air travel are constantly changing. If you have never availed yourself of the highly sought-after benefit of Space-A air travel or if you have several Space-A trips under your belt, you can learn a great deal from this book. This book can save you a lot of money in your air travels.

This *Military Space-A Air Basic Training™ and Reader Trip Reports* book will be kept simple. We are using a "step-by-step" approach which will cover, by example, all of the essential elements of Space-A air travel. This basic training will give you all the information that you will need to plan and successfully complete a typical Space-A air trip. We recommend that you, as the sponsor, study the ***Military Space-A Air Basic Training™ and Reader Trip Reports*** book, and that each dependent family member traveling with you study with you in order to sweep away some of the unknown. The information in this book will instill confidence in military members and their families while traveling Space-A.

SECTION I

AN IMAGINARY OVERSEAS SPACE-A AIR TRIP WITH THE FAMILY

TRIP PLANNING

TYPICAL SPACE-A AIR TRIP: For the purpose of ***Military Space-A Air Basic Training™ and Reader Trip Reports,*** this typical Space-A air trip is made by **a sponsor (Active or Retired), his/her dependent spouse, and two dependent children up to 21 years of age (or 23 years of age if a full time college student with a valid military dependent ID card (DD Form - 1173))** . Note: The sponsor may take some or all of his/her eligible dependents.

Some readers may prefer the term **"family members" to "dependents"** (we will discuss the documents required for Space-A travel later) but "dependents" is a necessary legal term in regard to the Space-A air travel regulation which must be met for Space-A air travel. (The DoD office of General Counsel has ruled that use of "Dependent" may be avoided, except to the extent necessary to satisfy explicit statutory requirements regarding entitlement to benefits and/or privileges.)

The typical trip begins on the East Coast of the Continental United States (CONUS) and continues to Western Europe and returns to CONUS; or it begins on the West Coast of CONUS and continues to the Western Pacific and returns to CONUS.

There are other typical variations, such as trips to the Middle East, Africa, Central and South America, the South Pacific (Australia/New Zealand), Alaska, and Hawaii. We cannot cover all possible alternatives in the limited space of this book. **The most popular Space-A air trip is from the CONUS East Coast to Western Europe and return** We have selected this trip, but with slight adjustments, you can adapt this scenario to your own travel needs.

Here we go on a Space-A air trip from the East Coast of the CONUS to Western Europe and return.

PLANNING YOUR SPACE-A TRIP: Your most important tasks to ensure success of your trip is **PRIOR TRIP PLANNING.** Most Space-A travelers have selected a destination or destinations for a variety of reasons which are clear, logical and rational for the travelers. Based on our scenario, we are going to Western Europe. Let us get more specific: What is our prime and most desired first destination in Western Europe? We want to make the Central Rhein River area of Germany our base of operation.

First we should consult Military Living's **Military Space-A Air Opportunities Air Route Map (Military Living's SPA Map),** which, along with its many other essential information features, is a prime route planning guide. It is indispensable to both the basic and advanced Space-A air traveler. This guide shows you, in graphic multi-color design, the scheduled air routes around the world. For convenience, this world map is centered on the continents of North and South America. From this air route map we can see that the best destinations in the German Rhein area are **Ramstein AB (RMS/EDAR), Germany, and Rhein-Main AB (FRF/EDAF), Germany . These three and four character letters/symbols are: the Federal Aviation Administration (FAA) Location Identifiers (LI) and the International Civil Aviation Organization (ICAO) location identifiers, which are fully explained in Appendix B. The popular FAA three character location identifiers are being replaced or supplemented more frequently by the ICAO four letter identifier for international stations in order to improve the positive identification of all stations worldwide. We will list these identifiers after each location/destination, when it is first mentioned, in order to reinforce the identifiers in your mind. You will find that these identifiers are indispensable to your understanding of Space-A air travel.**

We also note that we can depart from a variety of East Coast stations (north to south); McGuire AFB (WRI/KWRI), NJ, Philadelphia IAP (PHL/KPHL), PA, Dover AFB (DOV/KDOV), DE, Andrews AFB (ADW/KADW), MD, Norfolk NAS (NGU/KNGU), VA, and Charleston IAP/AFB (CHS/KCHS), SC.

The key question at this point in our trip planning is: From which CONUS East Coast locations can we depart in order to reach our desired destination in Germany? The Location Identifiers and Cross-reference Index, **Appendix C** of **Military Living's Space-A Air Opportunities Around the World** book also gives departure locations (other than the East Coast) for reaching the German Rhein River area. As you can see from **Military Living's Military Space-A Air Opportunities Air Route Map** and **Military Space-A Air Opportunities Around the World** book, there are more opportunities in terms of the number of flights to our destination **from Dover AFB, DE,** than the other departure locations. Do not rule out these other departure locations because they all have potential for flights to our selected destination. In fact, as you will see later, armed with a Space-A tip from **Military Living Publications,** we are going to register or apply for Space-A travel from at least four of the above listed departure locations/stations in order to increase our chances of selection for a Space-A flight. The Dover AFB, DE, departure location is feasible for us since we live in the Mid-Atlantic states area (Falls Church, VA, only 7 miles from Washington, DC). *Please note: The location identifiers and Cross-Reference Index, **Appendix C** of **Military Space-A Air Opportunities Around the World** book gives other departure locations for reaching the German Rhein River area.

BACK-UP PLANS: If making this trip is very important to you, you will want to have back-up plans to insure that you reach your goal or destination and return within your desired time frame. You cannot rely on DoD for this assurance. As the DoD Space-A directive says: ***"DoD cannot guarantee seats to Space-A passengers and is not obligated to continue travel or return Space-A passengers to the original point of travel."***

"How much is economy, round trip, double occupancy fare?"

If you must return by a specific date, you may want to purchase back-up commercial tickets. The commercial tickets will assure that you will be able to return within your desired time frame. If you do not use the tickets, there is a cancellation fee of $50 (at the time we went to press) for each leg of the trip. If for example, you purchase tickets to fly from Washington, D.C. or Philadelphia, PA to Germany and return, and you do not use the tickets, a cancellation fee would be charged. The terms and conditions of commercial air travel tickets change frequently as market place supply and demand, among other factors, change. Even with this cancellation fee, the savings of flying Space-A is far more than the cost of flying round trip commercially.

Unlike military Space-A travel, dependents/family members using commercial airlines may fly without their sponsor accompanying them. Another advantage of making commercial backup-up plans is that you do have a **reserved** seat on a flight heading for a definite location. If you have left your car at a particular base or airport, or if you need to return to the CONUS by a certain date, you can rest easy. Commercial back-up assures you that you can return to your point of departure in a timely fashion. Also, should you fall in love with a place and want to stay an extra day, week, month or even up to a year of the date of purchase of the tickets, you can do so with commercial back-up tickets. You will need to call the airline involved and see if a seat is available on the day you wish to travel.

Military airfares are available from most major cities throughout the United States to Germany. The tickets are usually good for one year from the date of purchase. Be sure and ask about this and other restrictions when buying a ticket, as rules are subject to change and may vary from airline to airline.

In **Military Living's R&R Space-A Report ®,** we often have one or more travel agency sponsors who are knowledgeable in locating a military fare in their computers, and who will work with you in the event you need a back-up ticket. Some agents do not want to work with anyone attempting to fly Space-A part or all of the way. Armed Services Vacations, a sponsor of our Washington area magazine, ***Military Living's Camaraderie Washington,*** has agreed to work with our military readers to help them utilize a combination of military Space-A and commercial travel. For info on back-up tickets, please call Godfrey Crowe, Colonel USA (Ret) at (703) 241-5911 (Toll free 1-800-658-8813). Godfrey may also be reached by mail at Armed Services Vacations, 130 Little Falls St., Suite B-100, Falls Church, .VA 22046. We hope that you will reward our sponsor with any other travel business you

may have, as there is not a lot of profit in issuing back-up tickets which may well be cancelled!

In addition, if in attempting to return home by Space-A you find yourself in a tight spot overseas you can check with **Scheduled Airline Ticket Office (SATO),** an organization which has ticket agencies on many U.S. military installations overseas. In addition some of our readers have reported that they were able to buy discounted one-way tickets from other commercial ticket agents located in the military air passenger terminals. We've heard that the closer you get to departure time, the cheaper the tickets seem to become!

This back-up travel arrangement does not use your Space-A air travel benefit, but it may save you untold money if you cannot for some reason use your Space-A privilege to meet your travel needs.

If you know of any other travel agents who would be interested in serving our military readership in this manner, please have them call Ann or R.J. Crawford at (703) 237-0203. We have 19 different publications which have advertising opportunities.

PRE-FLIGHT

REGISTRATION/SIGN-IN AT TERMINAL (STATION/AERIAL PORT): You, as the sponsor or lead traveler of a group, must register at the Space-A passenger departure or service center at each station from which you seek to depart on your Space-A trip. Registration can be made via FAX, mail/courier or in person. Registration times may be limited at some terminal/stations due to manpower limitations and demonstrated need. Please check **Military Living's Military Space-A Air Opportunities Around the World (SPA Book)** and reconfirm registration times with the terminal from which you plan to register for Space-A air travel.

Sponsor or Lead Traveler Registration. The Sponsor, or Lead Traveler, of a group may register for dependents and other persons who are traveling with them by sending a FAX, letter/courier application, or by presenting all required documents when registering in person. The sponsor must, if registering for Space-A travel via FAX, letter, or military courier (commercial couriers such as Federal Express and UPS were <u>not</u> acceptable at press time), provide the following: **Active Duty** - Application and service leave or pass form; a statement that boarder clearance documents are current; and a list of five countries. We recommend that you take all documents of each traveler in your party when registering in person, including each traveler's Uniformed Services ID card. There may be some inconvenience, but play it safe--take the ID card, as one terminal may sign up your dependents without your showing their card, but the other terminals may not!

"I'm signed up for England, Italy, and Iceland, but I'm not going anywhere. Since retiring I just enjoy getting on a roster knowing that I don't have to go."

DOCUMENTS

For our trip we need: Armed Forces Identification card, DD Form 2, (Green) for the Active Duty Sponsor or Armed Forces Identification Card, DD Form 2, Retired (Gray or Blue) for the Retired Sponsor; Uniformed Services Identification & Privilege Card, DD Form 1173 for all dependents accompanying Active Duty & Retired Uniformed Services members; passports for accompanying dependents and Retired Sponsors (Active duty for some destinations); and immunization records (recommended but not normally required).

Visas must be obtained when required by the **DoD Foreign Clearance Guides (See Appendix B, Personnel Entrance Requirements, Military Living's Space-A Book or check with the Space-A desk at any DoD international departure location/station or Military Personnel Offices which issue international travel orders.)** Please note that these personnel entrance requirements to foreign countries may be different (usually more restrictive) if entering the foreign country from a U.S. military owned or contract aircraft. Visas are not required for Western European countries but most Eastern European countries require them. See **Appendix R**: Visa Information.

Also, if you are Active Duty, you must have valid leave or pass orders in writing and **be in a leave or pass status throughout the registration, waiting, and**

travel periods. The DoD directive no longer requires that Active Duty personnel travel in the class A or B uniform of their service (the Marine Corps requires its' personnel to be in uniform). At press time, the uniform is not required for active duty personnel (and reservists in CONUS and U.S. Possessions) Some special flights from non-AMC terminals may still require uniforms. Please call and verify the uniform requirement of the terminal from which you will depart.

APPLICATION FOR SPACE-A AIR TRAVEL

As an Active or Retired member of one of the seven Uniformed Services (U.S. Army, U.S. Navy, U.S. Marine Corps, U.S. Coast Guard, U.S. Air Force, U.S. Public Health Service, and U.S. National Oceanographic and Atmospheric Administration), you can be the sponsor or lead traveler of a group, as shown in our imaginary trip. You may apply for Space-A travel via FAX, letter/courier or in person to the Dover AFB, DE, Space-A desk/counter at building 500, Passenger Terminal (24 hours daily), Telephone: Commercial 302-677-2854, FAX 302-677-2919, Defense Switched Network (DSN) 312-445-2854, Dover AFB, DE 19902-5496. From the main gate on 13th Street turn right on to Atlantic Avenue and continue. The Passenger Terminal is on the left in the Air Traffic Control (ATC) building (Bldg 500). **Directions to over 300 other terminals are listed in Military Living's Space-A Air Opportunities Around the World** book.

Present your documentation and request an application for Space-A air travel. **You will be given an APPLICATION FOR AIR TRAVEL, also known as the SPACE-A PASSENGER BOOKING CARD, Air Mobility Command (AMC) Form 53,** to complete. Please see a copy at **Appendix D.** You will be allowed to specify a maximum of five countries as destinations. The fifth designation may be "All" to take advantage of airlift opportunities to other countries that may fit your plans. All destinations within a country are included in this definition, i.e. "Germany" would include all destinations/stations in Germany. You will be assigned a category for your travel (based on the DoDD). In our example, Active Duty is category 2B (transportation priority D) and Retired is Category 4 (transportation priority R). Complete Categories, transportation priorities, documents required for each, and important restrictions/limitations are presented in table format in **Section IV.**

The DoD directive states: **"The order in which the categories are listed will be the normal precedence of movement. Eligible persons in each category will be furnished Space-A transportation on a first-in, first-out basis".** This policy is implemented by establishing at each station a Space-A roster of categories of travelers and their transportation priorities arranged higher to lower by their registration/sign-in times. **These times are recorded in Julian Date and Time**

format which is explained in Appendix C. This system allows registered Space-A persons to compete for seats within categories based on their date and time of registration.

"Hi. How's the waiting list to New Zealand?"

The Space-A desk clerk will verify the information contained in your application, examine your identification and travel documents for validity, and if you are an active duty sponsor, your leave or pass orders will be stamped on the back with the terminal name and the Julian Date and time at which you applied for Space-A air travel. **This category of traveler (Active Duty on Ordinary Leave) must have a validated copy of the leave or pass document when he/she is processed for the Space-A trip. If the application for Space-A is made by FAX or mail/courier the leave or pass must be transmitted with the FAX and will be effective NLT the date of the FAX or receipt of mail or military courier.**

Your application for Space-A travel will be entered into a computer-based waiting list by category, transportation priority, Julian date and time, and destination. Our application is entered for **Colonel (0-6) Francis O. Hardknuckle, USA (Ret)** and family, for four seats, Category 4, Transportation Priority R, for the following maximum five countries which we have specified: Germany, Italy, Spain, United Kingdom and "All".

If you are Active Duty, your application is entered for four seats, Category 2B, Transportation priority D, and for the same five countries. Your application will remain current in the computer, whether or not you report for Space-A calls, for a maximum of 45 days or until the end of your leave or pass, whichever comes first.

If you are Retired, your application will remain current in the computer, whether or not you report for Space-A calls, for a maximum of 45 days.

You will compete for seats within categories and transportation priorities that you have been assigned, based on the Julian date and the time of your registration. **Reservations for Space-A air travel will not be accepted, as there is no guaranteed space for passengers in any category.** If applying in person, as evidence of our registration, we will be given a Space-A Travel Information slip (developed locally) which will contain the five locations we have selected, our Category and Travel Priority, Julian Date and Time (accepted), Leave Expiration Date (Active Duty only), Sponsors name, Rank/Grade, Number of Seats Required, and the names of each dependent traveler. **The three dependents accompanying Colonel Hardknuckle are: Elizabeth Ann Hardknuckle, spouse; Linda Sue Hardknuckle, daughter, age 22, full-time graduate student; and Loren A. Hardknuckle, daughter, age 20, full-time college student. We will need this information anytime we inquire about our registration for Space-A air travel.**

"I have one seat left on this flight."

As we indicated earlier, in order to maximize our chances for obtaining Space-A air travel to our destination, we are also going to register at several other convenient Space-A departure locations. Why take this action? **Space-A seats are always a function of the number of perspective passengers waiting for a specific destination, the number of aircraft missions, the carrying capacity (available seats) on each mission, and the time at which all of these variables converge. Sounds complicated? True. But with so many unknown variables, your chances**

are improved if you register for air travel in the Space-A system at four points rather than one.

Through careful study of this text you can better understand and use the Military Space-A air travel system. We suggest that you also register for Space-A air travel at: Andrews AFB (ADW/KADW), MD, Philadelphia IAP (PHL/KPHL), PA, and McGuire AFB (WRI/KWRI), NJ. You can drive a car to accomplish this task, or there are several micro-bus transportation companies that service all of these stations for about $25-35 for each leg of the journey. Please see **Military Living's Military Space-A Air Opportunities Around the World** book for complete information on ground transportation at each terminal/station. For the strong at heart, you could also register at Norfolk NAS (NGU/KNGU), VA, to further improve your chances of obtaining a Space-A air trip to Western Europe.

TRAVEL DATES: In our planning we should establish approximate desired dates for departure. In addition to when active duty sponsors can obtain ordinary leave, we can establish an approximate departure date by examining several other pieces of information. First, check the Space-A backlog for the desired destinations at the departure terminals through the recording numbers or by telephoning the terminal personnel (**for these numbers consult your copy of Military Living's Military Space-A Air Opportunities Around the World book or Military Living's Military Space-A Air Opportunities Air Route Map**). Also check the number and frequency of flights by consulting the above documents. You must decide how much of your leave time you want to use while waiting for your Space-A flight; if Retired, your time is also valuable.

TRAVEL READY

Note carefully! If you are in or near the passenger terminal and you report for each scheduled and non-scheduled flight departure your chances of obtaining a Space-A flight to your desired destination are greatly improved. You have the right to stand-by for any flight that you believe you may have a reasonable opportunity on which to travel. In fact, you may stand-by for any flight, regardless of your chances or circumstances.

Many prospective Space-A travelers call the various terminals through the recording number, if available, or speak with the passenger terminal personnel to determine their numerical position on the waiting list for their registered destination. They also check their category and transportation priority on the waiting list, as well as the waiting list of higher category personnel. They do not go to the terminal to wait for a flight until their names are near the top of the Space-A waiting list and there is a flight scheduled in the immediate future. **This is a useful technique but**

don't forget that there are flights of opportunity, and if you are not in the passenger terminal and "travel ready", (discussed later), you don't fly.

The passenger terminal will not call you or notify you of a flight. In fact, they are prohibited from doing so by regulations. As a personal note, Ann, our author/ publisher, and I were in the Travis AFB (SUU/KSUU), CA, passenger terminal recently. There were 600 plus Space-A applications for Hickam AFB (HIK/PHIK), HI , and after repeated announcements in the terminal there were only 9 passengers who reported for the Space-A call and took the C-5/A flight with 76 seats to Hawaii. This Space-A air opportunity is repeated many times over every day around the world. To wait or not to wait in the terminal, or nearby, to keep you finger on the pulse of activity should be **your informed choice.**

SPACE-A FLIGHT CALL/SELECTION PROCESS: If you make the decision to wait in the terminal, or at any time you report to the terminal for the purpose of standing-by for a prospective flight, you should be **TRAVEL READY.** Travel ready simply means that you are ready and prepared to travel with very short notice. You have all of your required documentation, baggage, funds, and you are ready to report for a Space-A flight call/selection process and subsequent boarding. We have heard from many of our **R&R Space-A Report®** readers that they have missed a flight opportunity because they did not have enough time to return a rental car, move their baggage to the terminal, or other essential chores. You must be **TRAVEL READY** in order to maximize your chances for flying as you have planned.

Whether you have been waiting in the terminal for a long time, or you have reported for a flight call based on information that a flight is scheduled for departure to your desired destination, the processing procedure is the same. Many Space-A passengers enter their selected departure terminal and are processed and boarded for their desired destination within an hour, if and only if, they are **TRAVEL READY!**

Space-A seats are normally identified as early as two to three hours or as late as 30 minutes prior to departure. The standard notice time (Space-A call) for International/Overseas Space-A flights is two hours. This is the same standard for international flights on commercial airlines. In fact, many commercial international flights now require up to three hours for processing. Always check with the passenger service center for the **Space-A "Show Time"** for the flight on which you expect to travel. **"Show Time"** is the time that the Space-A Call/Selection begins and subsequent flight processing and boarding.

"Show Time" is important for the Space-A traveler because it not only means the time at which selection of registered Space-A prospective passengers begins; but

you must "now" be present to answer when you name is called in your category and transportation priority.

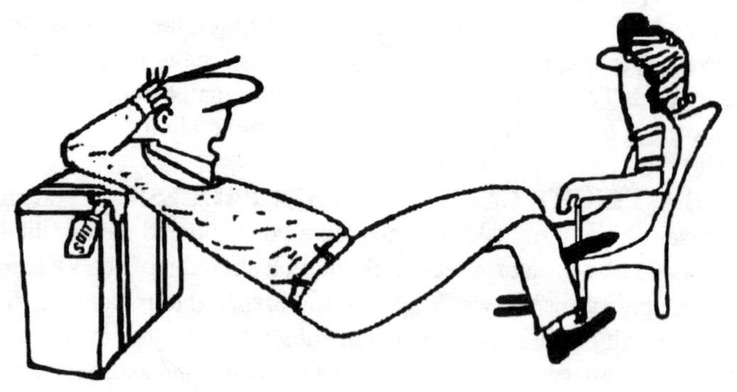

"I think I'm getting a terminal illness."

You will hear the announcement familiar to the seasoned Space-A traveler. **The Space-A Flight Call: "Anyone desiring Space-A air transportation to a series of stations - e.g., Rhein-Main AB (FRF/EDAF), GE. Please assemble at the Space-A desk or other location in the terminal." This is a roll call of prospective Space-A travelers.** The Space-A terminal flight processing team will be armed with the latest printout of the station Space-A roster (In fact, there is a Space-A roster for each category by date/time of sign up for ease in processing.) They will first call all the names of Category 1: Emergency Travel, Transportation Priority A, and process these travelers first. Next they will proceed to call travelers in each travel category, sub-category and transportation priority (from highest to lowest) by earliest Julian Date and time until they have filled all of the available seats for this flight.

The flight processing team knows the number of seats available and will select a sign-up Julian Date and time near the middle of the register. They will ask if anyone has that date and time or an earlier date and time, and either work forward or move back to the top of the list as appropriate. **Remember that the number of seats your party requires is a factor in your selection.** There must be sufficientseats for your party or the processors will move on and select someone with a later sign-up date, who requires fewer seats. As you can see, single and couple travelers may have a better chance of selection for Space-A air travel.

FLIGHT PROCESSING

Let us assume that your party has been selected for Space-A air travel and you are told to proceed immediately to flight processing (This may particularly be the case when the flight is non-scheduled). Therefore, you must be **TRAVEL READY** when you report for the Space-A call. It is not uncommon to proceed immediately to flight processing after the end of the Space-A call. Also, you can be told to report back for flight processing at a later time. In our case, we are told to report immediately to the Space-A desks for processing.

"You don't have to thank them for calling your name, George."

SPACE-A PROCESSING FEE: You may have heard that Space-A is "free". This is true. The Space-A $10 processing fee was eliminated in early 1993 by the Chairman of the Joint Chiefs of Staff, General Colin Powell.

Please note that all passengers departing CONUS, Alaska and Hawaii on contract commercial aircraft must pay a $6 head tax that goes toward airport improvements. Also, all Space-A passengers departing on contract missions from overseas to the United States must pay a $5 immigration inspection fee, a $5 customs inspection fee, and a $2 agriculture inspection fee.

PAYMENT OF FEES AND DOCUMENT REVIEW: You must pay these fees in United States dollars or via your personal dollar check. No charge/debit cards, travelers checks, or other payment instruments are accepted. These charges may be collected at the beginning or at other times in the flight processing.

The next action is an identification and travel documents check. The processing team will ensure that you have the following documents: appropriate ID cards for all travelers and that they are current; passports (signed) with required visas for retired service members and all dependents, and that they are current and valid; leave or pass orders for Active Duty personnel which were effective at the time of initial registration by the terminal and will be in force until at least the end of the flight for which you have been selected.

We have not been asked to show our **International Certificates of Vaccination and our Personal Health History, both as approved by the World Health Organization, Form: PHS-731 or similar (see Appendix H)** for each passenger, but we have them as recommended for use as needed. At some point, **the processing team will want to observe each passenger and verify their identity against their travel documents.**

BAGGAGE PROCESSING: You will of course need to pack prudently, sparingly, lightly and appropriately for your planned activities, and for the climate of Western Europe during the season in which you will be traveling. Each person is authorized two pieces of checked baggage, and each piece must not exceed 62 linear inches (length + height + width = 62 inches) in size and not more than 70 pounds, for a total of 140 pounds for both pieces. Family members and travel groups may pool their baggage allowances. **A Baggage Identification Tag (DD Form 1839 or AMC Form 20) is available for use on your hand-carried bags/items. Baggage claim tag USAF Form 94 (or similar) will be attached to each bag checked and you will receive a stub copy**. Sample copies are at **Appendix I.**

Space-A passengers may check oversized (golf clubs, snow skis, folding bicycles, etc.) baggage if it is the only piece to be checked (per person) and meets the weight requirement of 140 pounds total. Hand carried baggage cannot exceed 45 linear inches (length + height + width = 45 inches) and must fit under the seat or in the overhead compartment, if available. **Space-A passengers cannot pay for baggage in excess of the allowed weight limit.**

A few precautions: you cannot pack or carry weapons of any type (including cutting instruments), ammunition or explosive devices, chemicals or aerosol cans, and matches or other incendiary devices. Note: There are amnesty containers at the entrance to most terminals or boarding areas where you may discard any prohibited items annonymously and without penalty. Also, you must remove the batteries from all electronic equipment in your checked baggage. You may keep the batteries in your checked baggage. **At boarding time, you may be required to undergo an inspection of your hand carried baggage, or in some cases, all of your baggage.**

"Will those passengers holding boarding passes on mission four zero two, please assemble in the base chapel for last rites."

DRESS: Although the wearing of uniforms by Active Duty and Reservists is no longer required (except for USMC personnel), every Space-A passenger must be appropriately dressed. For your comfort and safety, you will want to dress with layers of clothes that can be added or removed as needed. Prohibited items are: open toed sandals and shoes (on military aircraft), T-shirts and tanktops as outer garments, shorts and revealing clothing. Strictly prohibited as an outer garment are clothing with profane language and any attire depicting the desecration of the U.S. Flag. If in doubt, call your air terminal for specific information. Use good conservative taste in your dress.

IN-FLIGHT MEALS: Since the Spring of 1988, AMC has developed and implemented a series of healthy-heart menus for their flights. These meals are prepared in their own in-flight kitchens. Snack menus cost from $1.50-2.00; which includes sandwich, salad or vegetable, fruit, milk or soft drink. Breakfast menus cost $1.50-2.00, which includes cereal or bagel, fruit, danish and milk or juice. Sandwich menus cost $2.50-3.00 and include sandwich, fruit, vegetable or salad, snack or dessert, milk, juice or soft drink. Expect in-flight meal prices to change. Also, prices are not standard worldwide. We reserve the sandwich menus and the breakfast menus since we will be most likely departing in the early evening (1900 hours local time) and arriving approximately 0800 hours local time the next day in Germany. If you reserve one or more meals, they will be entered on your **AMC BOARDING PASS/TICKET/RECEIPT, AMC FORM 148/2 (see Appendix G)** or similar form and your payment will be collected (cash or personal check). **You may often bring your own snacks aboard; however, you should check with the Air Passenger**

Terminal representative regarding any restrictions that may be in place due to Department of Agriculture regulations. No alcoholic beverages for consumption on military aircraft are allowed. Meals are free only on AMC contract flights. **Wine and beer are for sale only on AMC contract commercial flights.** Specialized meals are made available for duty passengers only, but you can request special food and passenger service personnel will help if they can. **If you need special food, we suggest that you bring your own in order to maintain flexibility. (See remarks above regarding snacks.)**

"I hope he orders an in-flight lunch."

BOARDING PASS/TICKET RECEIPT: At this point, we have successfully passed the flight call/selection process. All of our travel documents have been reviewed and accepted for travel; we have registered for in-flight meals and paid the charge; we have checked our baggage and received our claim checks; we have not selected a smoking section, since smoking on all military aircraft flights (including both military and contract civilian flights) has been eliminated; and lastly, we have been issued our Boarding Pass/Ticket Receipt which assigns our seats, verifies and documents the above items, and shows our one-time sign-up Julian date.

Now we wait for our flight number and destination to be called for boarding. They are listed on our Boarding Pass and the Terminal Departure board as follows:

MISSION	EQUIPMENT	DEPARTURE
AQB07TF	C-005A/B	105/94/1850L (2350G)

ROUTING	ARRIVAL
KWRI-KDOV-EDAF-LTAG	106/94/0745L (O645G)

Due to space limitations, we will not decipher the mission numbers in this elementary text. The equipment is a C-005A/B which is described in **Appendix L.** The departure date of 105 is Friday, 15 April, year 1994, 6:50 PM local time which is also 11:50 PM in GMT (Greenwich, United Kingdom). See **Appendices C and F** for a detailed explanation of the above. The routing is: McGuire AFB (WRI/KWRI), NJ, where the flight originated, Dover AFB (DOV/KDOV), DE, from which we are departing, to Rhein-Main AB (FRF/EDAF), GE, and on to Incirlick Apt (ADA/LTAG), TU where the mission turns around and returns to Rhein-Main AB, GE, to McGuire AFB, NJ. See **Appendix B** for explanation of Location Identifiers. Our arrival is 16 April, 1994 at 0745 local time at Rhein-MainAB, GE, which is also 0645 GMT.

"I've heard of getting bumpoed before, but this is ridiculous."

BUMPING: All Space-A travelers share a common fear - that they will be "bumped". The "bumping action" means that after being manifested (accepted) at a departure location for transportation on a flight, or at any station en route to your final destination, you are removed from the flight to make space for space required cargo and passengers.

Note that you cannot be "bumped" by another Space-A passenger except Category 1, Emergency Travel personnel, and then only with the direction of local authority. The good news is that if you are "bumped", you are placed on the Space-A list at the location where you are bumped **with the sign-up time, date and category that you received at your originating destination**. (This information is included on the manifest.) As a practical matter, "bumping" is very infrequent, because passenger service personnel take every action possible (reorganize cargo or move cargo to other flights) to preclude "bumping" one or more Space-A passengers.

PARKING: Since we will be about three weeks on our trip, we will want to park our car in the long term parking lot at Dover AFB, DE. The major portion of long-term parking is to your right as you walk in to the front of the terminal. Please ask the Space-A desk for directions or a map, and procedures. The location is a short distance away (very near the passenger terminal) and the parking fee is approximately $5 per week. There is an honor system for payment; lock the vehicle and keep the keys. Security police will patrol the parking lot often, thereby providing security for your vehicle. This is another example of how wonderful it is to be a member of the Military Family. Note: Do not leave your car parked in the short-term lot, it will be impounded and you will pay a fine to get your car back.

"They've sure tighened security since the last time I was here."

PASSENGER SECURITY SCREENING: Boarding has been announced as 1750 hours local time on Friday. We will be flying directly/non-stop from Dover AFB, DE, to Rhein-Main AB, GE, on a **C-5B Galaxy**. Our scheduled departure time will be 1850 hours local time, and we are scheduled to arrive at Rhein-Main AB, GE, at 0745 hours local time on Saturday morning. Similar to commercial airlines, **passengers are screened through electronic gates. Body searches may also be required**. All carry-on baggage, briefcases, purses, packages, etc. are screened before you are allowed to enter the secure boarding area. This security is for everyone's protection.

BOARDING/GATES: Space-A boarding processes are similar to commercial airlines. There is only one passenger class, and with the exception of some sensible and conventional practices, all passengers are boarded alike. Families with infants and small children are boarded first, passengers requiring assistance are next, then Distinguished Visitor/Very Important Persons (DV/VIP) passengers (Colonel USA, USMC, USAF; Captain USN, USCG, NOAA; Director USPHS), in pay grade O6+ and Senior Enlisted Personnel (Pay grade E-9) of the Military Services are boarded. **Colonel (O6) Francis O. Hardknuckle, USA (Ret), is the ranking officer on today's flight, so our party will be boarded next.** Then there is a general boarding of the remaining passengers. (**Note: Some terminals may not follow the DV/VIP ranking officer boarding and deplaning procedures.**)

Gates: There may be a few conventional and some not-so conventional obstacles to overcome. Today we are flying a **C-5B Galaxy aircraft**. We leave the boarding gate and enter a **"blue"** Air Force bus for a short ride to the aircraft parked on the ramp near the terminal. Today, courtesy stairs are not available. These stairs, when available, will allow easy access to the second deck passenger compartment of the C-5B Galaxy aircraft. We are instructed to walk through the forward cargo door of the aircraft, and up a cargo ramp to a point near the rear of the cargo department. **There we will climb (not walk) up an almost vertical (approximately 18 feet), metal ladder with hand rails to the C-5B passenger flight deck.** Female passengers should preferably wear slacks (more about travel dress later).

The passenger processing team and the aircraft crew will provide you with assistance. Elderly people, mothers with infants in arms and people with physical handicaps will have difficulty in climbing the passenger ladder. However, Space-A regulations require that you must be capable of boarding and exiting the aircraft with limited assistance unless you are **a retired 100% Disabled American Veteran (DAV) and traveling with an assistant entitled to Space-A air travel within his or her own right. (Carefully note that 100% DAV's have the Space-A privilege if and only if they are retired and are in possession of a grey or blue ID card, DD Form 2.)**

If you have a nonapparent handicap such as a hearing impairment, asthma, pacemaker, etc., please advise the passenger processing team at time of check-in. Frequently the C-5B and similar aircraft will have mobile stairways (stair truck) for direct access to the passenger deck, but as indicated earlier, no stairs are available for today's flight departure. **This is probably the most difficult boarding encounter that you will find in your Space-A travel experiences.**

"Hi. What have you got from Edwards to Travis next month?"

IN-FLIGHT

SEATING: The seats in these aircraft are conventional commercial airline seats. However, we immediately note that the seats face to the rear of the aircraft. This configuration is for safety purposes. The only difference you will notice from commercial airline seating, which faces forward, is a different sensation during take-off and landing of the aircraft. Pillows, blankets, and other comfort items are available. We hope you have brought along your own reading or other amusement materials (games) as there are no reading materials, in-flight movies or music or military aircraft.

CLOTHING: Your clothing should be loose fitting and in layers. Wea comfortable walking shoes. Women should preferably wear slacks and a blouse o sweater. Take along a light jacket or coat depending upon the climate. Remembe that western Europe is always a bit cooler than the corresponding latitudes in Nort America. The layers of clothes will come in handy if you return, as a passenger no a patient, on a Medical Evacuation (MEDEVAC) flight where the temperatures ar kept quite warm for the comfort of the patients.

REST ROOMS: The rest rooms on the C-5A/B are adequate in number an are similar to commercial airlines. They are unisex and you are expected to kee them clean after your use.

CLIMATE CONTROL: As mentioned, the climate in the cabin is similar to that of commercial airlines. The temperature can vary and you have been given blankets to make your travel comfortable. If it gets too cold or warm, you may ask the cabin personnel to adjust the temperature. Whether or not your wishes are complied with may depend on the capabilities of the aircraft, the crew and other passengers' needs. The floor of the passenger cabin/deck of the C-5A/B tends to be cooler than is comfortable for some passengers.

NOISE: This is not a noisy aircraft in the passenger cabin. You may experience some higher than normal levels of noise during take-off, landing, and special maneuvers and turns. At other times you should be most comfortable. There is no need for ear plugs on this flight (this is not true of some other flights, such as those aboard the C-130E aircraft). If you travel often on noisy aircraft, you should obtain form fitting ear plugs at your medical facility ear clinic. These plugs will improve your comfort and conserve your hearing. The crew may have wax self forming ear plugs. Ask for a set if you would be more comfortable.

"Airman Smith will demonstrate how to use the life preserver quick release. Fortunately we do not anticipate an in-flight emergency."

SAFETY: You will be required to use your safety belts as instructed by the aircraft commander or cabin crew. Only walk about the aircraft when allowed to do so. Never tamper with controls, doors or equipment within the aircraft. Listen very carefully to the in-flight safety lecture given by the cabin crew members prior to takeoff. Pay particular attention to the location of exits and identify the exit which is most convenient to your seating. Become familiar with the location of your emergency oxygen supply and life preserver since most of this flight will be over water.

ELECTRONIC DEVICES: The playing and usage of radios, recorders, TV,s computers, etc., which may interfere with aircraft navigation, RADAR and radio systems is prohibited. As noted earlier, these items in your checked baggage must have the batteries removed.

SMOKING: Smoking is not allowed on United States military aircraft or military contract flights.

REFRESHMENTS: There is always fresh water, coffee and tea available in the galley. The meals we ordered will be served at the appropriate times, given flight conditions. Our dinner sandwich meal will be served about an hour and a half after departure and our breakfast meal will be served approximately one hour before our arrival.

Our flight departed as scheduled and all travel en route has been flawless. Further, we are told that we will be arriving at Rhein-Main AB, GE, as scheduled at 0745 hours, Saturday morning. Wow! All this flying for "free".

POST-FLIGHT

ARRIVAL OF AIRCRAFT/CLEARANCE: We are notified that our flight will arrive on time. We are given immigration and customs forms to be completed

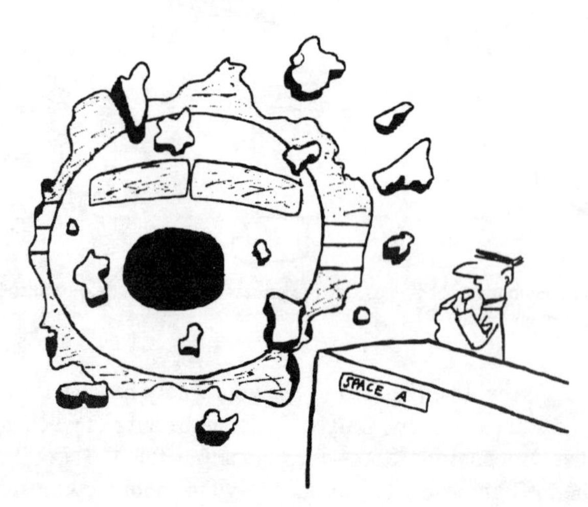

"Air Mobility Command announces the immediate arrival of mission one six eight. Passengers will be in the terminal in approximately two minutes."

for each passport and/or ID card holder. The flight attendants spray fumigation/insect repellant throughout the cabin as required by international health rule. After landing, our manifest and declaration of the health of the crew and passengers will be handed over to German immigration authorities, or the United States authorities acting on their behalf. This is necessary to obtain clearance for the crew and passengers of our aircraft to disembark. Once clearance has been obtained, deplaning will begin as instructed.

DEPLANING: Prior to deplaning, we will be given instructions regarding post-flight processing such as Immigration, Baggage Claim and Customs. We're fortunate today in that we'll be deplaning through passenger level doors onto a mobile passageway into the air passenger terminal at Rhein-Main AB, GE. **Colonel (O6) Hardknuckle USA (Ret)**, is the ranking officer on our flight. He and his family will depart the aircraft first, followed by a priority for deplaning designated by the flight cabin crew. In most cases, those families with children and other people in need of assistance during deplaning will be the last to exit. **(Note: Some terminals may no longer follow the DV/VIP/ranking officer boarding and deplaning procedures.)**

IMMIGRATION: Next we must report to the immigration processing station/desk, and present to the German/U.S. authorities our immigration/customs form and other travel documents. Sponsors will need their ID card. Retired sponsors will also require a passport. Dependent family members will need their ID cards and passports. All passengers must be present for this processing to verify their identities. The authorities will examine your travel documents carefully. As part of the processing, they may also check for persons barred entry and other wanted persons. Passengers may be questioned about previous visits to Germany, the purpose of their visit, places to be visited, and accommodations (place of residence) in the country.

We are delighted to find that all is in order for our family; our immigration forms and passports are stamped, and we are instructed to claim our baggage and report for customs clearance. We are lucky, our four bags are waiting for us on the carrousel. Colonel Hardknuckle has allowed us only one bag, that can be rolled or carried, per person. We move with ease to Customs Inspection.

CUSTOMS: The customs inspection is a breeze. We are not bringing any dutiable or prohibited items such as tobacco, coffee, tea, or alcoholic beverages into the country. Our customs documents are stamped, and we are now cleared to leave the terminal. Before we leave for our vacation, a very important task regarding our return Space-A trip should have our top attention.

REGISTRATION FOR RETURN SPACE-A AIR OPPORTUNITIES: We should register for departure back to CONUS as soon as possible (**it is a good idea for the sponsor, with everyone's travel documents, to sign-up for return airtransportation while other members fetch taxis or attend to other arrival chores**) from Rhein-Main AB, (RMS/EDAR) GE, if this is our planned return departure terminal. We can also register at Ramstein AB, GE, or for that matter, at any other terminal in Europe. **A special caution! If you register to return to CONUS from any European terminal (i.e. Rhein-Main AB, GE), and subsequently take a local flight from Rhein-Main AB, GE (for example, a visit to Aviano AB (AVB/LIPA), IT, your registration to CONUS and any other registrations from Rhein-Main AB, GE will be deleted from the computer based registration.** Note: Under the new remote sign-up procedures Colonel Hardknuckle

"Do you hear me George? I said take out those stupid ear plugs until we get on the plane!"

could have signed-up in advance by FAX for his party's return air travel prior to arriving at Rhein-Main AB. This rule is in the Space-A DoD Directive, and applies at any station worldwide. Once you depart on any flight from a station, all of your other applications for flights from that station are deleted from that station's registration system.

RETURN FLIGHT TO CONUS

As told by Linda Sue Hardknuckle, daughter: During our trip, we took the "blue" Air Force bus from Rhein-Main AB, GE, (check with terminal personnel) to Ramstein AB, GE. We enjoyed dinner in the Officers' Club and a comfortable night in two adjacent rooms in the Temporary Military Lodging (TML) facility, bldg 305 Washington Avenue, tel 011-49-637-44589/42589. We had signed up immediately upon arrival at Ramstein AB, GE for flights to Italy and Spain in order to retain our Dover AFB, DE initial registration time and travel priority. The priority can be retained and is effective for a trip which continues in the same general direction from CONUS. In this case, we are continuing southeast to Spain/Italy. The countries must be on our CONUS sign-up and if active duty, we must be on leave.

SIDE TRIPS: We were fortunate to get a flight early the next morning to Torrejon de Ardoz AB (TOJ/LETO), SP, near Madrid, Spain. On arrival **we were careful to obtain a "Letter of Entrada and A Base Pass" from passenger service personnel before leaving the base, to insure our easy reentry**. This is a Spanish AB where the USAF is a tenant under our Status of Forces Treaty with Spain. We also signed-up for a flight to San Guisto Airport (PSA/LIRP) in Pisa, Italy which we took after our four-day visit to Madrid.

The flight to Pisa, was aboard a C-130E aircraft with bucket seats along the side and cargo in the middle of the aircraft. The aircraft was very noisy, particularly on takeoff and on landing/braking. The crew chief passed out **formable wax ear plugs to dampen the noise.** The ride was short and it was a new and interesting experience. After two days of shopping in Pisa, we departed Pisa, for Rhein-Main AB, Germany, on a **C-009A Nightingale aircraft, MEDEVAC flight**, which was a special experience for my sister Loren, who is studying nursing.

At Rhein-Main AB, GE we checked our progress on the waiting list for CONUS. Our names had only moved up the Category 4 waiting list from number 404 when signing-up, to number 314. However, our plan was to leave in approximately one more week's time. So we continued with our travel plans.

After a week visiting the Rhein and Mosel river areas via a major company rental car, we returned to Rhein-Main AB, GE, to discover that we were number 273 on the waiting list for CONUS.

There were two flights scheduled for the next day. The first was a flight to RAF Mildenhall (MHZ/EGUN), UK, and then on to Charleston AFB, (CHS/KCHS) SC. The second scheduled flight was to go directly to Dover AFB, (DOV/KDOV) DE,where we had left our car, and was exactly where we wanted to go. We stood-by for the Space-A call for both flights but there were a lot of emergency leave persons in Category 1, some Active Duty leave persons in Category 2B, a few military personnel on permissive no/cost TDY in Category 3, and only 10 Retirees in Category 4, who were ahead of us on the list. Some of these Retirees obtained Space-A transportation. We did not!

CIRCUITOUS ROUTING: After this failure, and armed with tips from **Military Living Publications,** we discovered a flight scheduled for the next morning which was routed from Rhein-Main AB, GE, to Incirlik Airport (Adana) (ADA/LTAG), TU. The flight was scheduled to depart Rhein-Main AB, GE, at 1015 hours local time and arrive at Incirik Airport (Adana), TU, at 1415 hours local time and remain at Adana, for a 15 hour crew rest after arrival. The schedule for this mission was to fly the next day from Incirlik Airport, TU, back to Rhein-Main AB, GE, where the aircraft would remain three hours on the ground and then continue on to our desired destination, Dover AFB, DE. If we can be manifested, we will be flying our favorite C-5B Galaxy aircraft. (*We are manifested, with a new sign-up date, as there are few active duty Space-A passengers on leave who are returning to duty at Incirlik Airport, TU.*)

"We're going to NBO by way of Ngu, Ton, and Siz."

We have learned another principal of Space-A Basic Training, namely: **that a straight route may not be the best route for Space-A Air Opportunity travel. Circuitous routing is often used to get to your destination on Space-A.** We are eager, risk-taking, Space-A travelers - so we go for it! **Please recall that our registration for CONUS from Rhein-Main AB, GE, will automatically be deleted from the system when we take the flight to Incirlik Airport, Adana, TU.**

Once we arrived at Incirlik Airport, Turkey and signed up for our return flight to Dover AFB, DE, we lose our original Rhein-Main AB, GE date/time sign-up and pick-up a new date/time at Incirlik since we were changing direction completely and heading back to CONUS. We remembered to list Dover AFB, DE, as our final destination. **Incirlik Airport was able to manifest us all the way to Dover AFB, DE, thereby allowing us to bypass having to terminate our flight at Rhein-Main AFB, Frankfurt GE and re-register for Dover AFB. Otherwise, we would have been back where we started before going to Incirlik.**

Gosh! The first leg of the trip worked as planned. We arrived at Incirlik Airport after about four-hours flying time, at 1415 hours local time. We departed the aircraft first. After showing our travel documents to Turkish and U.S. Officials we moved smoothly through immigration, claimed our bags and cleared customs. Colonel Hardknuckle, USA (Ret) registered our group for the return to CONUS. We were 20-23 on the Category 4 list for CONUS. We concluded that we had a very good chance of obtaining travel on this flight, which was scheduled to depart at 0445 hours the next day. This flight had an early Show Time of 0245 hours.

"If we can't get out of here by tomorrow, you take commercial to Seattle and Tokyo, and I'll meet you sometime next month in Atsugi."

We had called from Rhein-Main AB, GE, after we were manifested to reserve our Temporary Military Lodging, so all was set. We took a taxi to our Temporary Lodging facility, bldg 1081 7th Street at Incirlik Airport. Again we found Incirlik Airport (the USAF is a tenant on a Turkish AB) to be a well-appointed facility, with a splendid Officers' Club for drinks and dinner, and a very comfortable Temporary Military Lodging facility.

The next morning we called the Space-A desk about one hour before the show time to discover that our flight would not be departing until 1700 hours due to changes in cargo and other space-required needs. There were no other flights to Rhein-Main AB, GE, or CONUS prior to this time. So on the bright side, we had most of the day to spend as we pleased. The new Show Time was 1500 hours but my dad, Colonel Hardknuckle, wanted to maximize our chances for travel and set our personnel Show Time at the terminal for 1400 hours.

In order to spend the day wisely and have some fun, different members of our family group planned the day's activities. Dad set up a day of tennis. Other family members arranged tours with the USAF Ticket and Tour Office of castle ruins and museums in the local area. A relaxing travel day was had by all.

We arrived at the passenger terminal promptly at 1400 hours. Our flight was posted as follows: Leave LTAG/ADA, 122/94/1700; Arrive EDAF/FRF, 122/94/2210. We knew from our **Space-A Air Basic Training** that we were scheduled to leave from Incirlik Airport (Adana), TU, at 5:00 PM local time on Monday, May 02, 1994.

RETURN FLIGHT: Luck and good planning were with us. The Space-A flight call was made and we were on the flight to Rhein-Main AB, GE. We had paid the $3 per person fee for the sandwich menus we ordered for this leg of the trip at Rhein-Main AB. Aware that we would have some seven and a half hours ground time at Rhein-Main AB, GE, before our flight continues on to Dover AFB, DE, we attempted to obtain billeting at Rhein-Main AB, GE, but none was available.

We were very fortunate to obtain Temporary Military Lodging (TML) space at the Patriot Inn in Jefferson Village, Darmstadt, GE, a short drive south of Rhein-Main AB on the E-4 Autobahn. One of our transportation options was to rent a car but, due to the late hour we were unable to. The other option was a taxi, which we took straight to the Patriot Inn. We rushed to bed because we had to be back at the airport for a 0400 hours Space-A call. **We were smart travelers - we brought our toilet articles, sleeping attire, and a clean shirt/blouse in our carry-on luggage, so we were prepared for the overnight in Darmstadt, GE, as our luggage was checked through to Dover AFB, DE.**

Being cautious types, we made arrangements with the desk clerk for a wake up call. All went well, and we met our early morning Space-A call. **Since we were continuing on the same flight on which we arrived, we checked in with Passenger Service and signed up and paid for the in-flight meals, and received our boarding passes.** We recognized some of the faces of Category 4 people who were still waiting for transportation to CONUS at Rhein-Main AB when we departed for Adana, TU. They met the Space-A call but were not boarded on our flight to CONUS due to lack of available seating. We enjoyed our trip to Turkey and were now on our way home. If we had been unable to obtain military lodging, we would have had to use, if available, a pensione (small hotel) in the local area.

ARRIVAL AND CLEARING: The flight to Dover AFB, DE, was uneventful. **Before landing we were given immigration and customs forms to complete as a family group.** We completed the immigration forms for our passports and ID cards as a family group. Likewise, we completed the customs declaration as returning residents and we made an oral declaration, since the value of goods which we brought into the country did not exceed $400 per person. **We were not bringing in tobacco, tea, liquor or any other exempted items. Also, we were not bringing in items for other persons.** We arrived on time at 123/94/1505L - also known as

" We went for broke. Quad fifties, twin forties, the works. Like I told ya yesterday, somebody's got to lead the troops. Remember Ike? That's what I told him. Know what he said? 'Go get'em soldier.' How about that?"

Tuesday, May 03, 1994, 3:05 PM. We checked through immigration, claimed our four bags, and cleared customs.

In a stroke of good luck, my dad, Colonel Hardknuckle, obtained a local ride to claim our station wagon located in the long-term overflow parking facility. He met a Space-A passenger who wanted to hear of his Space-A experiences, and who still had his car at the main terminal.

We met one of our friends, LtCol Zachary Sadbag, USAF (Ret) whose bag did not arrive with his flight. As a matter of interest, we inquired as to how this loss is handled. He told us that the Dover AFB, DE, Passenger Service Personnel were most helpful. The procedure was similar to passengers who have lost baggage on commercial airlines. **He was requested to fill out a AMC Form 134, BAGGAGE IRREGULARITY REPORT. A copy of his baggage claim check (USAF Form 94) was secured to this form. He was also requested to fill out a AMC Form 70, RUSH BAGGAGE MANIFEST specifying whether he wanted them to hold the bag for pick-up when it came in or forward it to his home in the Washington, DC, area at his expense.** LtCol Sadbag elected to wait for his bag which was scheduled to arrive on the very next flight from Germany in about four hours.

POST SPACE-A TRIP REPORT: We departed the terminal for a short 2 and 3/4 hour trip via car to our home in Falls Church, VA. On the way home we compared notes on our trip which are to be submitted the next day to our dad, who will be preparing an **After Trip Report.** It will emphasize the Space-A lessons we have learned, and will be circulated among our friends. We will also send it to **Military Living's R&R Space-A Report** ® for possible publication in their Reader Report Sharing section. This information will be very useful to us and others on future trips.

Linda Sue Hardknuckle

- NOTES -

SECTION II

AN IMAGINARY SPACE-A AIR TRIP BY CONUS MEDICAL EVACUATION (MEDEVAC) FLIGHT

BACKGROUND

MEDICAL EVACUATION (MEDEVAC) FLIGHTS: These flights are flown all over the world in support of the medical programs of the Uniformed Services. The vital United States Air Force, Air Mobility Command (AMC) aeromedical airlift mission is implemented by highly trained medical technicians, flight nurses, and aircraft crews which moved nearly 80,000 DoD patients on some 4,500, C-9, C-141, and C-130 missions during 1993. This mission includes both litter and ambulatory patients. In addition to these patients, **thousands of Space-A Air Opportunity passengers were also served by the MEDEVAC system worldwide.**

THE MEDEVAC SYSTEM: There are several elements of the worldwide MEDEVAC system which operates in CONUS. Next, each major overseas theater of operations, i.e., Europe and Pacific, has its own internal and intra-theater MEDEVAC system. Lastly, there is a MEDEVAC system which operates between CONUS and overseas theaters. We have discussed Space-A travel between CONUS and overseas with return, and travel in the European theater. In this section we will only discuss Space-A air opportunities in CONUS via MEDEVAC missions. **As our example we will use an Active Duty Uniformed Service Member, First Lieutenant John A. Hardcharger, USAF, stationed in Northern California.**

CONUS MEDEVAC FLIGHT EQUIPMENT: The major aircraft used in CONUS MEDEVAC is appropriately named the **C-009A/C NIGHTINGALE.** This aircraft has been derived from the DC-9, Series 30 Commercial airliner. The C-009A/C has been converted in to an aeromedical airlift transport. This aircraft in this configuration has been in service in the USAF since August 1968. The aircraft accommodates a crew of three, medical staff of five, 40 litter patients or 40 ambulatory patients, or a combination of both. More details regarding the specifications and performance of this aircraft are at **Appendix L. Two other aircraft the C-130E Hercules and the C-141A/B Starlifter, have also been configured for aeromedical airlift and are used for long distance Theater and Intra-Theater aeromedical airlift missions.** Specifications, Configuration and Performance of these aircraft are also at **Appendix L.**

WHAT IS DIFFERENT ABOUT SPACE-A ON MEDEVAC FLIGHTS? With the help of **1Lt Hardcharger**, we will take another one of those favorite

Space-A air opportunity trips. First, let's examine the special aspects of Space-A travel on aeromedical airlift - commonly know as MEDEVAC - flights. The seating for Space-A passengers on the C-009A/C Nightingale MEDEVAC flights is commercial airline type, reclining chairs facing toward the rear of the aircraft. **The C-009A/C is a very smooth-riding aircraft. However, for patient comfort, the temperature is kept at approximately 75 degrees and may be kept even higher. Caution: wear clothes in layers that you can take off or add as needed. The temperature will only be adjusted for the comfort of patients.**

" Air Mobility Command announces the immediate loading and departure of flight number one to Kitty Hawk."

The Space-A seating is away from the litter patients, frequently in the forward part of the passenger cabin. Litters patient are boarded first, followed by ambulatory patients, Space-Required (if any), and Space-A passengers. A key item to remember is that MEDEVAC flights can be diverted from their planned flight schedule, without notice, at any time, to pick up or discharge MEDEVAC patients.

PLANNING YOUR CONUS MEDEVAC FLIGHT: The CONUS MEDEVAC system is **based on a monthly proforma schedule which includes many stations.** These stations are visited on different days of the month depending upon the stations' need for MEDEVAC (Aeromedical) Services. For specific details, please see: **Military Space-A Air Opportunities Around The World Air Route Map (Military Living's Space-A Map) and Military Space-A Air Opportunities Around the World (Military Living's Space-A Book).** With these valuable tools in hand, 1Lt Hardcharger is ready to take his Space-A air trip.

As told by 1Lt Hardcharger: My plan is to travel from Northern California (San Francisco, CA) area, where I am stationed, to the East Coast of the United States (Washington, DC), then to Florida (Orlando, FL, area) and return to the West Coast (San Francisco, CA). I examine the MEDEVAC schedule carefully and determine that my Space-A MEDEVAC travel plan is, indeed, feasible.

I note from **Military Living's Space-A Book and Space-A Map** that I can take a flight from Travis, AFB (SUU/KSUU), CA, on Sunday to Andrews AFB (ADW/KADW), MD . The purpose of my leave travel is to visit relatives, see some fun places, and be introduced by cousins to their friends. **I have my Air Force leave form signed by Colonel Michael Q. Topmissile, USAF, my Project Manager, and it is valid for 20 days.** I register at Travis AFB, CA by Telefax (FAX) the day that my leave is effective only for CONUS destinations. This means that I am eligible for any flight that departs from Travis AFB, CA, to a CONUS destination. It is reassuring to know that I can travel on any mission, on any Services' aircraft.

REGISTRATION AND DEPARTURE: I have registered for Space-A travel on Saturday, my first day of leave, at Travis, AFB, CA, for CONUS. I have a registration for one person, Category 2B: Ordinary Leave, CONUS destination, with my Julian Date and Time of Registration from my Telefax (FAX) transmission receipt (**see Appendix C** for an explanation). I report on Sunday for a Space-A call for Andrews AFB, MD. Fortunately, I am selected for the flight. Wow! I am on my way and it is free. I reserve the Snack menu (healthy heart menus) for $1.50 which includes a deli sandwich, salad or vegetable, fruit, and milk or a soft drink.

ROUTING AND REMAIN OVERNIGHT (LODGING): I had called ahead to check on Temporary Military Lodging (TML) at Andrews AFB, MD using **Military Living's Temporary Military Lodging Around The World book.** Since I am in a leave status, lodging at Andrews AFB is on a Space-A basis. I decide to take my chances. The flight to Andrews AFB, MD, is smooth and uneventful. Immediately upon arrival, I register in person for Space-A Air to CONUS. Next, I claim my bag in the Passenger Terminal, building 1245. Tel: 301-981-1854. ((Note AMC is making an effort to use the same line number (extension 1854) as the Air Passenger Terminal Information number at all AMC bases. This may take some time to accomplish.))

Due to the late hour of arrival, the shuttle bus is not operating. I have two ways to reach the billets: walk the approximately one mile, or call for a commercial taxi to come on base. In the interest of saving money and since I have only one small bag, I decide to walk. It pays to travel light. You will have more flying options when traveling in CONUS if you keep your baggage under 30 pounds. The baggage limit on the smaller executive aircraft is 30 pounds.

I arrive at the TML and check-in at building 1374 Arkansas Road which is a 24 hour per day operation. I am fortunate to obtain a room for the night. I have dinner at the Officers' Club in building 1352, which is a short walk from the TML.

"I wanted to be a pilot, but my mother wouldn't let me."

I leave on Monday from Andrews AFB, MD for Keesler AFB (BIX/KBIX), MS. **The flight to Keesler AFB, MS, is diverted from our scheduled route to pick up an emergency MEDEVAC patient at Moody AFB (VAD/KVAD), GA and this makes it a very long day for everyone on board.** After arrival, I register in person for CONUS (MacDill AFB (KMCF/KMCF), FL) and claim my bag in the Passenger Terminal, building 0223. I hitch a ride with a SMSgt, who is stationed on base, to the TML check-in at building 2101 where I obtain a room for the night.

I join a friend for a short ride to Biloxi, Mississippi where we have dinner and return to the base. The next day (Tuesday) I take a flight from Keesler AFB, MS, to MacDill AFB, FL. On arrival at MacDill AFB, Passenger Terminal, hangar no. 4, **I claim my baggage and register in person for CONUS so that when I am ready to return, I will be much higher on the Space-A waiting list.**

I reserved a rental car by calling ahead to Tel: 813-840-2303, located at Dayton Avenue on base. My rental car is ready and I depart to spend a few wonderful days visiting relatives and enjoying Florida sights.

After a week in Florida, I am ready to head home to California. I take the Tuesday flight from MacDill AFB, FL to Keesler AFB, MS, where I remain overnight. On Wednesday I take a flight from Keesler AFB, MS, to Andrews AFB, MD. On Thursday I take a flight from Andrews AFB, MD via Kelley AFB (SKF/KSKF), TX to Travis AFB, CA, and home.

THE SCHEDULED MEDEVAC SYSTEM: Careful planning and coordination was required to reach my desired locations. The MEDEVAC system has been very reliable and has met my needs. The requirement to stop overnight several times was a different experience for me but very enjoyable. **This is a wonderful system for the unaccompanied individual Service member, Active or Retired, to move about CONUS.**

Given a short period of time you can go to almost any area in CONUS and return to your station with minimum waiting time. Please also note that while you are registered for CONUS at a departure location, you can take any flight regardless of the type mission to any location in CONUS. **If you are patient, relaxed and flexible, any Space-A trip is easy.** If I had missed a leg on my flight schedule, I would have checked to see about other Space-A flight opportunities. Space-A air travel is not always a series of flights in a direct line. **It was a comfort to know that if all my Space-A plans "fell through," I could use my active duty military ID card to get a big discount on a commercial flight.** Most military bases have travel offices, known as Scheduled Airline Ticket Offices (SATO), or you can call the airline direct.

1Lt John A. Hardcharger

- NOTES -

SECTION III

AN IMAGINARY SPACE-A AIR TRIP BY GUARD AND RESERVE PERSONNEL

GUARD AND RESERVE PERSONNEL ELIGIBLE FOR SPACE-A AIR TRAVEL

GUARD AND RESERVE PERSONNEL: All Guard and Reserve personnel are eligible for Space-A air transportation, with the exception of those who are not receiving pay or have not completed their requirements for retirement. First, let us look at the seven Uniformed Services to see which Services have Guard and Reserve components. There are Army and Air Force National Guard components; but there are no National Guard components in the other five Uniformed Services. There are active Reserve components in the Army, Navy, Marine Corps, Coast Guard and Air Force. There are no Active Reserve components in the USPHS and NOAA.

ACTIVE DUTY STATUS RESERVE COMPONENT MEMBERS: In order to be classified as an Active Duty status Reserve component member, you must be a full time member of a National Guard or Reserve Unit or drill independently and receive Guard/Reserve pay. This means that you drill or train with your Guard or Reserve unit on a regular basis, you are in Active status and you receive pay for your attendance at drills or training sessions. You may be an officer, warrant officer or enlisted grade. This describes the Active Duty Status Reservist of all the Uniformed Services (Armed Services).

RETIRED RESERVISTS (GRAY AREA): Guard and Reserve members of the Uniformed Services (Armed Services) cannot receive full retirement status until they attain age 60. **When Reserve component personnel, who have not attained age 60, receive their official notification of retirement eligibility, they may continue to travel as Reservists.** The date that reservists receive official notice of retirement eligibility until they reach the age of 60 and are officially retired is know as the "Grey Area". When the Retired Reservist reaches age 60, he or she is fully Retired and receives the DD Form 2, Blue ID card and begins to receive retired pay. At this point the Retired Reservist has all the benefits of any Retired member of the Uniformed Services.

DOCUMENTATION REQUIRED FOR GUARD AND RESERVE COMPONENT PERSONNEL TO FLY SPACE-A: First, Reserve component personnel must have their DD Form 2 Red ID card. Second, they must have a copy of **DD Form 1853, AUTHENTICATION OF RESERVE STATUS FOR TRAVEL ELIGIBILITY, signed by their unit commander within the past 180**

days. Gray Area Retirees, must have a copy of their official notification of retirement eligibility. **A copy of DD Form 1853 is not required for Gray Area Retirees.**

GUARD AND RESERVE TRAVEL-GEOGRAPHY AND DEPENDENTS: Guard and Reserve component members may travel in CONUS and between Alaska, Hawaii, Puerto Rico, the U.S. Virgin Islands, American Samoa and Guam (Guam and American Samoa travelers may fly via Hawaii). **Except for Guard and Reserve Retirees aged 60 and above, Guard and Reserve personnel may not travel Space-A to a foreign country. The above geographic restrictions on reserve component personnel travel also apply to the Gray Area Reserve component Retiree.**

Guard and Reserve component personnel cannot be accompanied by dependents (as a minor exception, when reserve component personnel are on Active Duty for training overseas, which is a regular Active Duty status, their dependents may accompany them on Space-A air travel within the overseas area). Gray Area Retirees cannot be accompanied by their dependents prior to attaining age 60 and receiving full retirement and their DD Form 2, Blue ID card. **At this point their dependents will also be eligible for the DD Form 1173 Uniform Service Identification and Privilege Card reflecting the retired status of the sponsor.**

THE SCOPE OF GUARD AND RESERVE SPACE-A TRAVEL: Although there is a restriction against Guard and Reserve personnel traveling to foreign countries, nonetheless, there are many exciting places to visit. Also, from these overseas (OCONUS) points, you can continue your travels to many foreign countries at a minimum expense. For example, American Samoa is three-fourths of the way to Australia/New Zealand. Likewise, Puerto Rico and the U.S. Virgin Islands open up to you the entire Caribbean area. Guam is on the doorstep of Asia. Japan, the Philippines, Taipei, the Pacific Islands and Indonesia are nearby. Of course, there are also the foreign countries bordering CONUS which you can easily visit.

TRIP PLANING

AN IMAGINARY SPACE-A AIR TRIP BY GUARD AND RESERVE PERSONNEL: This Space-A trip will be taken by **Sergeant (SGT), E-5. Roy L. Straitlace, USANG**, DC National Guard (who lives in Washington, DC), and **Staff Sergeant (SSgt), E-5, Kathleen L. Truelove, USAF Reserve**, Pope AFB, NC (who live in Fayetteville, NC). They met during Operation Desert Shield/Storm at a United Services Organization (USO) canteen and have stayed in close contact since that time. Now they have planned a Space-A air trip to Puerto Rico where they will have a joint vacation and hopefully get to know each other better under more normal conditions.

ROUTING: Sergeant Straitlace checks **Military Living's Space-A Map and Book** and finds that there are almost daily flights from the **Washington Naval Air Facility (NAF) (NSF/KNSF), DC, located at Andrews AFB, MD to Norfolk NAS (NGU/KNGU), VA.** His review of Space-A air opportunities reveals that there are also numerous flights from **Norfolk NAS, VA, to Roosevelt Roads NAS (NRR/TJNR), PR.** Knowing the value of prior planning, SGT Straitlace also checks return flights because he must return within 10 days to his civilian job as a Federal Security Guard at a DoD agency in Washington, DC. He notes that there is a MEDEVAC flight from Roosevelt Roads NAS, PR, through Norfolk NAS, VA, to Andrews AFB, MD, each Sunday. Also, there are numerous flights from Roosevelt Roads NAS, PR, to Norfolk NAS, VA. To reconfirm the non-scheduled Navy link of his planned route from Washington NAF, DC, to Norfolk NAS, VA, and return, **SGT Straitlace calls the Washington NAF, Building 3198, Tel: 301-981-2740-2744, where he speaks with PO1 Norman Scuppers, USN who assures SGT Straitlace that there are frequent flights with numerous Space-A seats on this route.** SGT Straitlace discovers that most of the aircraft on this route are the following types: C-009 A/C Nightingale, C-21A Executive Aircraft, C-130 A-H Hercules, and P-3C-Orion.

Staff Sergeant Truelove, using Military Living's SPA Book and Map, finds that there are many flights each week from Pope AFB (POB/KPOB), NC, to Norfolk NAS, VA which continue on to Roosevelt Roads NAS, PR. Most of these flights are via C-130E mixed passenger/cargo missions. As the result of her active duty in the Persian Gulf area, SSgt Truelove has her own form-fitted ear plugs and also has some experience flying the C-130E aircraft. At SGT Straitlace's insistence, she checks the return flights because she, too, must be back in 14 days to report for a new job that she has obtained as a medical technician at an area hospital. Luck is with everyone, and there are many return flights from Roosevelt Roads, NAS, PR to Pope AFB, NC through Norfolk NAS, VA or through Charleston AFB (CHS/KCHS), SC .

APPLICATION FOR SPACE-A TRAVEL: The two sergeants apply for air travel at their respective locations. SGT Straitlace applies in person for air travel at both the Washington NAF, DC, and at Andrews AFB, MD. SGT Straitlace has learned from **Military Living's SPA Book that Space-A seats not used by the NAF are offered to Andrews AFB, and vice versa.** SGT Straitlace applies for travel to CONUS in order to cover any destinations that he may need, and for Puerto Rico and the U.S. Virgin Islands.

SSgt Truelove applies in person for air travel at Pope AFB, NC. She applies for travel to CONUS, Puerto Rico and the U.S. Virgin Islands.

Each sergeant also sends a Telefax (FAX) with a copy of their Authenticated DD Form 1853, Authentication of Reserve Status for Travel Eligibility, to air passenger terminal Roosevelt Roads NAS (NRR/TJNR), PR for registration for return travel to CONUS. This will give them a higher position on the Space-A register than if they wait until they are in Puerto Rico to apply for return Space-A air travel to CONUS.

THE OUTBOUND TRIP

Our sergeants are all set to go. It looks like Thursday is the best day to travel from Norfolk NAS, VA, to Roosevelt Roads NAS, PR. SGT Straitlace reports to the Washington NAF, DC, early on Wednesday for four flights that will be departing during the day for Norfolk NAS, VA. SGT Straitlace obtained the above flight departure information by calling the Washington NAF after 2200 hours on Tuesday. **An early morning flight is his on a C-009 which delivers him to the Norfolk NAS, VA, before 1000 hours on Wednesday. Before leaving the terminal, he registers for Puerto Rico and is told that there will be two flights on Thursday.**

OVERNIGHT: SGT Straitlace obtains billeting at the enlisted billeting office, building 1-A, Pocahontas and Bacon Streets, Tel: 804-444-4425 for one night. He calls SGT Truelove from the NCO Club, where he has dinner and tells her about his trip to Norfolk, VA. SSgt Truelove plans to report for both flights at Pope AFB, NC to Norfolk VA. Of course, SGT Straitlace will be at the Norfolk NAS to meet her.

MEETING: SSgt Truelove is not on the first flight, a C-130E. Meanwhile, a flight to Roosevelt Roads NAS, PR, arrives. SGT Straitlace, who as a good Space-A traveler follows the rule of **"see flight, take it," does, and he arrives in Puerto Rico before she does!** SSgt Truelove does make the second flight, another C130-E, and continues on to Puerto Rico. SGT Straitlace is waiting to take them to San Juan, about 50 miles northwest, by bus. **Before leaving the airport, they both confirm that the passenger terminal has received their FAX application for return flights to CONUS and that their requests have been entered into the Space-A register.**

VACATION: They have reserved adjoining rooms for the first three nights in a San Juan hotel which grants special military rates. After they become oriented to the island, they plan several trips to the interior rain forest and beautiful beaches. **They have an enjoyable vacation and discover a greater than-expected renewal of their friendship. In that vein, they also plan their engagement announcement.**

RETURN FLIGHT

The two sergeants keep checking back with the Space-A desk at the Roosevelt Roads NAS and find that there is a flight one week from the following Tuesday. When they report to the passenger terminal "travel ready" they find that their names are near the top of the waiting list. They are both selected for the flight, which is another C130-E. The flight is routed from Roosevelt Roads NAS, to Norfolk NAS, VA, and on to Pope AFB, NC. SGT Straitlace departs the flight at Norfolk NAS, VA. He registers for a flight which departs the next day to CONUS, and remains overnight at the Naval Base BEQ. **The next morning, Wednesday, SGT Straitlace catches a C-21 flight directly to the Washington NAF, DC. He was able to take this flight on a small executive aircraft because his luggage did not weigh more than 30 pounds.**

REFLECTIONS: Guard and Reserve members are not allowed to fly Space-A to foreign counties until they reach age 60 and are Retired; however, they can have lots of fun flying by Space-A in CONUS and to and from United States Possessions overseas, such as Puerto Rico, "the star of the Caribbean."

- NOTES -

SECTION IV
SPACE-A ELIGIBILITY AND DOCUMENTATION REQUIREMENTS TABLES

SPACE-A ELIGIBLE PASSENGERS (Note 1)	(1)	(2)	(3)	(4)	(5)	(6)	(7)	(8)
TRAVEL CATEGORIES & TRANSPORTATION PRIORITIES	Armed Forces ID Card, DD Form 2 (Green), DD Form 2-NOAA (Green), or PHS 1866-1 (Green).	US Uniformed Services Retired ID Card, DD Form 2 (Gray or Blue), or Uniformed Services ID Card, DD Form 2-NOAA (Gray or Blue).	Armed Forces ID Card, DD Form 2 (Red) AD Status Reserve Component.	Uniformed Services ID and Privilege Card, DD Form 1173	Valid Leave Orders/ Authorization, Travel Orders or Transportation Authorization (reference to paragraph of regulation used to authorize travel).	Authentication of Reserve Status for travel eligibility, completed DD Form 1853, or official notification of retirement eligibility,--Age 60 not attained.	Valid United States (or other country) Passport with appropriate Visas. Tourist Card only, if sufficient to enter some countries.	I. International Certificates of Vaccination, and II. Personal Health History as appropriate.
TRANSPORTATION OF SPACE-A PASSENGERS BETWEEN CONUS AND OVERSEAS (OUTSIDE CONUS) CATEGORY 1: Emergency Travel. (TP-A)	NOTE 2: Active Duty Member.	NOTE 2: Red Cross verified emergency.	NOTE 2: Red Cross verified emergency.	Dependents of uniformed service member, DoD civilian employee overseas; Non-Command sponsored, No return to overseas duty station of sponsor.	Sponsors, emergency leave orders (including dependents).	Required with column (3), (Note 2).	Dependents of Active Duty, DoD civilian employees, and others authorized Space-A travel overseas.	As appropriate
CATEGORY 1-HF: Hostile Fire. (TP-B)	Active Duty Member.	N/A	N/A	N/A	EML Leave Orders from HF area to EML destination and return.	N/A	As required, Armed Forces ID Cards only req to most EML dest.	As appropriate
CATEGORY 2-A: Environmental, Morale Leave Program (EML). (TP-C)	Active Duty Member	N/A	N/A	Civilian sponsors & all dependents to include USO & DoDDS (on holiday).	EML leave orders (included dependents).	N/A	Dependents of Active Duty, DoD civilian employee & other authorized Space-A overseas.	As appropriate

CATEGORY 2-B: Ordinary leave, other. (TP-D)	Active Duty Member	N/A. Col (4) cont: Foreign exchange military personnel in DoD position; US uniformed service personnel on conventional leave; Close blood, or affinitive relative on PCS w/sponsor.	N/A. Col (4) cont: Dep of AD, exchange mil, (Note 3) civilian DoD & others auth SPA. Dependents and TDY/TDA is covered in Note 8. Cadets, Midshipmen, and foreign Cadets.	US civilian patients in recovery & non- medical attendant. Medal of Honor Recip & accom dep. Military pers on non-charged leave for house hunting w/one dep. Note 3. Continued in Col (2 & 3).	Leave orders or approved pass. Medal of Honor ID & travel card, as needed.	N/A	Dependents of Active Duty, DoD civilian employees & others authorized Space-A overseas.	As appropriate.
CATEGORY 2-C: Unaccompanied EML program. (TP-E)	N/A	N/A	N/A	Unaccom dep age 17+ w/o adult (-17: adult req). DoDDS teachers & their dep on summer break.	EML program, leave orders required for all travelers.	N/A	All dependents of sponsor, as required.	As appropriate.
CATEGORY 3: Students, dependents, permissive TDY, etc. (TP-F)	Active Duty Member	N/A Col (4) cont: Dep students of AD, DoD civ & ARC attending secondary, trade, and undergraduate schools under age 23: 1 round trip/yr. Dep of AD & DoD civ emp who becomes eligible for transportation.	N/A Col (4) cont: AD personnel on Permissive/no-cost TDY orders. Dep acquired overseas: return to CONUS, AK & HI only. Dep travel for service academy testing.	US Govt agency employees & Dep travel fro medical treatment/consultation, only on dedicated Ambassadorial support flights. Continued in Col (2 & 3).	Student travel orders, Dep joining sponsor orders, Permissive TDY orders, Acquired dependents orders, Academy testing orders, US employees medical orders.	N/A	Dependents of Active Duty, DoD civilian employee and others authorized Space-A overseas.	As appropriate
CATEGORY 4: Reserve Components, retired, other. (TP-R)	N/A	Retired members	Note 4. Active Duty reservists, gray area reservists, senior level ROTC cadets.	Dependents of retiree, dep of AD, DoD civilian employee & ARC, to return to CONUS for enlistment in a uniformed service.	Note 5. PMS&T letter.	Active Duty Reservists and gray area reservists.	Dependents of Active Duty, DoD civilian employee and others authorized Space-A overseas.	As appropriate

TRANSPORTATION OF SPACE-A PASSENGERS WITHIN CONUS CATEGORY 1: Emergency Travel. (TP-A)	Active Duty Member	N/A	N/A	N/A	Emergency leave orders	N/A	N/A	N/A
CATEGORY 2: Ordinary leave, others. (TP-C)	Active Duty Member	N/A	N/A	For exchange military in DoD position. Spouse, children, dep parents of MIA or POW w/Chief of Svc approval. Medal of Honor recipient. AD non-charged leave for house hunting.	Leave orders or approved pass. Medal of Honor ID & travel card. Letter from appropriate military dept for dependent of MIA & POW.	N/A	N/A	N/A
CATEGORY 3: Non-chargeable leave. (TP-E)	Active Duty Member	N/A	N/A	N/A	Non-charged leave orders.	N/A	N/A	N/A
CATEGORY 4: Reserve Components, Retired. (TP-R)	N/A	Retired members.	Active Duty Reservists, Gray Area Reservists.	Note 5. No dependents in CONUS.	Note 5.	Active Duty Reservists, Gray Area Reservists.	N/A	N/A
TRANSPORTATION OF SPACE-A PASSENGERS WITHIN & BETWEEN OVERSEAS AREAS CATEGORY 1: Emergency travel. (TP-A)	Active Duty Member	Note 2: Red Cross verified emergency.	NOTE 2: Red Cross verified emergency.	DoD civ emp overseas, command sponsored dependent, non-command sponsored dep to terminal nearest emergency, but no return to overseas station. ARC & other civ overseas. Foreign exchange military pers in DoD. Foreign emp of DOD to home country & return.	Emergency leave orders.	N/A	Dependents of Active Duty, DoD civilian employees and others authorized Space-A travel overseas.	As appropriate
CATEGORY 1-HF: Hostile Fire. (TP-B)	Active Duty Member	N/A	N/A	N/A	EML leave orders.	N/A	As required, Armed Forces ID cards only req to most EML Destination.	As appropriate.
CATEGORY 2-A: EML program (TP-C)	Active Duty Member	N/A	N/A	Accompanying dependents including: USO professionals, DoDDS teachers (on school holidays).	EML leave orders	N/A	Dependents of Active Duty, DoD civilian employees & others	As appropriate

CATEGORY 2-B: Ordinary leave and other. (TP-D)	Active Duty Member	N/A	N/A	Foreign exchange military personnel in DoD positions & dependents. US uniformed svc patient. Close blood or affn rel, DoD civ emp, ARC overseas w/PCS of sponsor. Foreign-born dep spouse & dep children overseas of MIA w/Chief of Svc approval. Dep of AD w/students up to age 23. Circuitous trvl of dep. Cadets & Midshipmen on leave. US civ armed forces patients. Medal of Honor recip & Deps. AD on Non-charged leave.	Leave orders, approved pass, travel authorization and service department letter for MIA families.	N/A	Dependents of Active Duty, DoD civilian employee, and others authorized Space-A overseas.	As appropriate.
CATEGORY 2-C: Unaccompanied EML. (TP-E)	N/A	N/A	N/A	Dep 17+ (-17 accom by adult EML eligible fam member), DoDDS teachers & dep summer vacation.	EML orders	N/A	Required for all travelers.	As appropriate.
CATEGORY 3: Students, Dependents, Permissive TDY, Other. (TP-F)	Active Duty Member	N/A	N/A	Dep secondary, trade, & college student of AD, DoD civ, and ARC; under 23, 1 round trip/yr. Dep approved to join sponsor overseas. AD on permissive no-cost TDY. Acquired dep trvl to new duty station of sponsor. Dep trvl for svc academy testing. US Govt civ emp trvl for medical evaluation/treatment (US Amb Spt Flts only).	Official orders for function to be performed and issued by the appropriate service department.	N/A	Dependents of Active Duty, DoD civilian employee, and others authorized Space- A overseas.	As appropriate
CATEGORY 4: Reserve Components, Retired. (TP-R)	N/A	Retired Member	Active Reservists, Gray Area Reservists, Senior level ROTC Cadets.	Active Duty dependents, Dep of AD, DoD civ emp, ARC return to US for enlistment in uniformed services.	Note 5 & 7.	Active Reservists, Gray Area Reservists.	Dependents of Active Duty, DoD civilian employees, and others authorized Space-A overseas.	As appropriate.

Section IV (Continued)

NOTES:

1. REGISTRATION FOR SPACE-A TRAVEL. Space-A rosters are maintained for those locations served by each base. To compete for Space-A, eligible persons must register via Fax, letter/military courier, or in person with all documentation as listed in Paragraph 4-2(a) through (h) in DoDD 4515.13-R (Draft) and summarized above. The lead traveler may register the entire family group as long as the required documentation is presented. Passengers may register for a maximum of five countries (or four countries plus all) or CONUS and need not compete for all seats offered. All passengers are automatically removed from the register after 45 consecutive days. All Space-A passengers dropped from the list may register again with a new date and time of sign-up.

2. Category 4 personnel may be upgraded to Category I with a Family Emergency validated by the Red Cross and approved by the local commander or his representative while overseas (outside CONUS).

3. This (Space-A) privilege may not be used for travel of the dependents to or from the sponsor's restricted (unaccompanied) tour duty station, or for travel to establish a home for dependents.

4. Members of Reserve Components traveling between AK, HI, PR, VI, AS and GU (GU & AS travelers may fly via HI) and the CONUS In the following circumstances: A. Active Duty status members. Members who have received official notification of retirement eligibility, but have not reached the mandatory retirement age of 60.

5. Reserve Officer Training Corps (ROTC) students of the Army, Navy, and Air Force receiving financial assistance, and those enrolled in advanced training, traveling in uniform during authorized absence from school. Students will present a document bearing the signature of the senior commissioned officer, who is the professor in charge of the ROTC program at the civilian educational institution. The document will identify the student, by name, as being in the advanced course or enrolled under the Financial Assistance Program.

6. Space-A travel of dependents in the CONUS is not authorized except as follows: Dependents may travel on the domestic leg segments of international (overseas) flights during the beginning or end of their international flight.

7. Active Duty Reserve members on pass who are on Active Duty in an overseas area for any length of time. tn addition to the required ID card these

members are required to present an Active Duty order authorizing them to be in the assigned overseas area during the period travel is requested.

8. When Active Duty members are authorized leave in conjunction with TDY/TAD, the dependents may not travel between the permanent duty station and the TDY/TAD location. However, upon the member's arrival at the TDY location, dependents are authorized Space-A travel with the member under the following conditions: A. When the member's permanent duty station and temporary duty station are within the CONUS, travel to an overseas area and return is authorized; B. When the members permanent duty station and temporary duty location are in different countries within the overseas area, travel from the temporary duty location to the CONUS or within the overseas area, other than to the location at which the member is permanently assigned, is authorized; C. When the member's permanent duty station and temporary duty location are within the same country, travel within the overseas area is authorized from, and return to, the temporary duty location; or from, and return to, a location other than the member's permanent duty station.

NEW SPACE-A TRAVEL CATEGORIES

In the very near future, DoD is expected to announce a redefinition and regrouping of the **Space-A Travel Categories and Transportation Priorities.** It is very important to note that, this redefinition will not alter or change the relative travel priorities of the existing eligible Space-A travelers. The new redefinition will, among other things, change the travel categories from four to six. Also a total of 44 travel groups will be identified. These groups are now identified in the present regulation. A major improvement envisioned by this change is that Travelers Status and Situation will be grouped into their respective geographical travel segments, i. e. origin and destination combinations are: CONUS to CONUS; Overseas to Overseas; CONUS to Overseas; Overseas to CONUS. The travel groups, their travel categories and geographical travel segments are expected to be formatted into a table which will clearly identify the travel status, situation and eligibility of all Space-A travelers. Watch Military Living's **R & R Space-A Report®** for this change when it is approved and published by DoD.

SECTION V

READER TRIP REPORTS

Many readers of **Military Living's R&R Space-A Report®** have told us of their successful Space-A trips. The following Reader Trip Reports will provide a guideline for you on using Space-A to various locations throughout the world. ***Please remember that prices, times, routing, and other details in Reader Trip Reports should be viewed as things that can, and do, change.***

ASIA

This letter is from a few years back, as our bases in the Philippines are now closed.

Dear Ann,

We made a trip to San Francisco for visa renewal. Stopped at Travis [AFB] and signed up [for Space-A]. Spent the holidays at home then returned to Travis, and got on a C-5 for Philippines. Made a two-day stop in Hawaii because of mechanical problems. Guam is not the best place to stay. On our two previous stops there were no quarters available and this time [it] was the same. We did arrive early enough to rent a car that we never took out of the terminal parking lot. We used it to sleep in which was not too bad with the seats put back. It beats sitting up in the terminal.

We had a 0100 show and 0355 departure for Clark AFB. We arrived three hours later and checked into the base hotel.

We called the terminal and were told a C-5 with 73 seats was going to Singapore with a 0100 rollcall. We made the show time only to find all the seats had been given to the Navy. This is just one of the things that can happen.

Checked the board and found a C-130 with 15 seats was due to land at 0530 so we decided to eat and stick around and see if we could get seats. We got lucky and 6 1/2 hours later we were in Bangkok. The C-130 was hot, noisy, and the strip seats were not the most comfortable, but who is to be unhappy when you find yourself in Bangkok with only a $20.00 cost for transportation? A military bus took us across the airfield to the civilian side where we took a limo to the Federal Hotel.

After returning to our hotel in Chaing Mai we decided to fly back [commercial] to Bangkok. After another three days in Bangkok we decided to see Indonesia. We got an excursion ticket on Singapore Airlines for Bangkok, Singapore, Jakarta and return.

We spent two days in Singapore where we called to see about Space-A back to Clark AFB. We were told that we shouldn't have any trouble getting seats. We stayed in the YMCA on Orchard Road in Singapore.

We spent four more days in Singapore and got Space-A back to Clark AFB. It took three days to get out of Clark on a C-5 back to Travis. We had an 18 hour crew rest in Guam. There were no quarters so [we] sat up in the terminal. We spent one night in Hawaii and back to Travis.

None of this would have been half as nice if it were not for the wealth of information we had gleaned from [y]our R&R Reports.

Our first Space-A trip to Asia was great. We were a little uptight not knowing just what to expect, but once really got us hooked. The second trip was fun all the way. If you are not going Space-A, you're missing something!!

Sincerely,
Harvey C Baker
Salinas, CA

AUSTRALIA

Dear Ann,

We have finally taken a chance on Space-A. We thought we'd try a tough one for the first, New Zealand and Australia, and all (the) others would be easy.

We flew commercial to Ontario, CA and stayed with a Korean War buddy who lived in San Bernardino. Four days later we were on our way to Hawaii/Hickam on an L1011 that had been chartered to take a Marine company to Kaneoe. It had a hundred open seats, and was only announced on a telephone tape at 8 PM, with a 7 AM departure, 5 AM roll call. We were "stuck" in Oahu for 5 days, and stayed on at (the) Hale Koa, the rest at the Hickam VOQ.

The flight through Pago Pago to Christchurch was uneventful, (the) C-141 chilly but fast. New Zealand should be a must for everyone. Great people...perfect sunshine, unbelievable scenery. But be sure to take lots of insect repellant for the blackflies.

We missed two departures, one that would have gone to Australia, the other direct back to Hickam. We were 32 and 33 on the list trying to get back stateside, and saw no way that we could make the next flight, since we had been told it had only 12 seats. We went to rollcall anyhow, and found that the small detachment at

Christchurch had spent the night consolidating the four half loaded pallets of freight onto two, providing 35 seats, and everyone got out!

Another uneventful flight to Hickam, where a 30 hour crew rest was scheduled. We watched two California-bound flights fill up with active duty people, and then, magic, a C-130 bound for Colorado Springs with beaucoup seats. We made it, and as I spotted the Springs just ahead, the plane turned north. The loadmaster came back to apologize that we had been diverted to Buckley ANGB because of crowded crew quarters at Peterson AFB at the Springs. They parked the plane at Buckley and bussed us all to Lowry AFB, Denver, where the crew and all others bunked in transient quarters, and we taxied ten miles home. What a finish. This spring we are going to try to make Norway.

Sincerely,
Leonard & Artie Horner
LtCol USAF (Ret)

The Horner's letter showsonce again how AMC personnel go the extra mile for Space-A travelers.

Dear Ann,

I called Norton Air Base Terminal in Southern California. The flights originate there. Two flights on different days each week. One flight goes to Christchurch, New Zealand and then to Richmond, Australia. The other goes to Richmond and then down to New Zealand. All go via Hickam Field in Hawaii and Samoa.

The actual flying time from Norton is between 19 and 20 hours in a C-141. This is not a flying Hilton but it gets you there FREE. Seats are fastened to the floor. They are upholstered and can be adjusted to a reclining position with plenty of room to stretch your legs. When aloft you may walk around as you choose. Many people sleep.

Toilet facilities are adequate. Men and women use the same. Blankets and pillows are available. Water and coffee may be had. Depending on the load, 10-25 Space-A people can be accommodated.

(Following an 18 day stay) I...caught the Space-A flight (from Christ Church) to Richmond Air Base 35 miles west of Sydney, Australia. Because of fog we skipped New Zealand and returned to Norton via Pago Pago and Hawaii.

(On AMC flights) time must be on your side in case of a layover but it is all certainly worth the effort and adjustments.

Chaplain Gomer Rees, COL, USA, (Ret)

Ed. note: Chaplain Rees was 83 when he made this trip. He is now deceased...did he ever love Space-A! Also, Norton is now closed.

AZORES

Dear Ann,

My wife and I left Minneapolis on May 20th on a National Guard C-130, headed for Torrejon, Spain, via McGuire AFB and the Azores.

We were bumped at McGuire. However, we rented a car for [$31.00] at the AMC terminal and drove to Atlantic City for a two hour visit to Trump Palace (The Taj Mahal) and Resorts International. We then returned to McGuire and stayed overnight for [$16.00].

We flew out the following evening on a C-141 jet cargo plane and were in the Azores five hours later the following morning. There is a four hour difference from the East Coast.

We obtained accommodations at the Azores Air Base at Family Billeting, renewing our stay each day for the five days we were in the Azores. The cost was [$20] per night for the two of us.

We rented a car at the Recreation Center for [$18.00] per day. It was delivered to the Recreation Center. We returned it at the AMC terminal where the car agency picked it up before we left.

The weather was cool, humid, and raining most of the time. Nevertheless, we took a group tour of the island in addition to driving around on our own the two other days.

Highlights were the many small farms and dairy cattle, lush foliage, and abundant lava rock... and many pot holes and cobble streets. Also, the natural swimming hole on the north end of the island, Duck Pond, and a delightful picnic area on the east end.

Sulphur springs and volcanic caves were other attractions which were cloud-covered the day we were there.

The major town of Angra is about 1/2 hour from the base. The group tour guide took us into the fort that commands the harbor. We had lunch at a very clean and reasonable restaurant, enjoyed the gardens in the park at the center of town, and had some espresso at a nearby bakery.

Base facilities within a block or two of family billeting included a 24-hour snack bar, NCO Club, Officers' Club, BX, Commissary, Recreation Center, and swimming pool.

The Azores are a jumping off place for Mildenhall, Rhein-Main, Torrejon, and Aviano-Sigonella-Athens-Adena, as well as the USA.

There were several flights back to the USA, including a contract DC-10 to Philadelphia and a C-141 to McGuire. We went on the return flight of the Minnesota National Guard C-130 to Minneapolis.

Yours truly,
Paul Bruening
CDR SC USN (Ret)
Bloomington, MN

Ed. note: Athens, Athenai Air Base is now closed.

ENGLAND

Dear Roy and Ann:

My wife and I returned on July 10th from a trip to England and Ireland; and I want to share our experiences with you. Your publications including the R & R eports were most helpful.

On June 13, 1994, we flew Space-A from St. Louis airport (having driven up from San Antonio) on a contract 747 (Tower Air) to Rhein-Main. At the St. Louis airport I was able to park the car [for] free at the Missouri Air National Guard 314-263-6301) which is about one block from the terminal building. They are used o having people leave cars there for several months and assured me it would be safe. The flight to Rhein-Main was only about half full. They have been going each Monday at least during June with empty seats. The flight was great: good food, movies.

We arrived at Rhein-Main at about 0800 on June 14th and I immediately igned up on the Space-A list for a flight to Mildenhall. I should have listed my final destination as England instead of Germany when I originally signed up at Scott [AFB] for the St. Louis departure. However, we were able to get on a contract 727 the next morning to Mildenhall, so nothing was lost. Billeting did not have any space in the Hotel which is not far from the passenger terminal, but we were able to get into Building 600 which is across the air strip in the family housing area, and it is on the shuttle bus route. We had a good dinner that night at the Officers' Club.

At 0515 the next morning we boarded the shuttle bus for a ride to the passenger terminal for a 0730 departure on the contract 727 to Mildenhall: a one-hour flight with an excellent breakfast; no charge for the flight. On arriving at Mildenhall, I signed up for Space-A return to the States. Our plans were to spend a couple of weeks driving around Ireland and then return to England for a week before heading home. At the SATO office (Mildenhall) we bought tickets for a flight that afternoon on Ryan Air from London Stanstead Airport to Dublin, Ireland. Since the trains were on a one day strike and the buses were fully booked, we used M & L Taxi (located on Mildenhall) for the forty minute drive to Stanstead. Stanstead is a modern, under-used airport with no crowds; it was a pleasure.

After an hour flight to Dublin, we rented a car at the Dublin airport. There are many auto rentals at the airport including the usual stateside ones. We shopped around and found that the price could vary $150 for the same type of auto. We got the best price from Avis with a free upgrade. Before leaving the airport, we had the Irish Tourist Board desk make us a reservation that night at a B & B at a seaside town not too far from the airport. We spent the next two weeks driving across Ireland to the northwest coast and then essentially following the coast south and east back to Dublin. We stayed entirely in B & Bs (room with bath) for the two of us including a large Irish breakfast (juice, cereal, eggs, bacon, toast, coffee). We drove approximately 150 miles a day visiting sites listed in material that we had obtained from the Irish Tourist Board before leaving the States. Each afternoon we located a B & B from the directory that we had also received from the Tourist Board. We found the B & B's to be excellent and the least expensive accommodations. The hosts were most friendly and helpful, often providing tea and cookies upon our arrival. We ate the evening meal in the pubs as the food was good and cheaper than in restaurants. All of the pubs, restaurants, gas stations, and a few B & B's accepted Visa/Mastercard (which provides the best exchange rate).

We hated to leave Ireland at the end of the two weeks, and [we] plan to return. The weather was cool to cold, mostly cloudy with occasional light rain (a welcome change from San Antonio). The return to Mildenhall was uneventful, again using M & L Taxi from Stanstead to Mildenhall. We arrived back at Mildenhall on June 30 and immediately started trying to get a flight back to the States. We stayed at Mildenhall for a week trying for a flight but to no avail. There were few flights originating there and not much traffic coming through. The flights that did come through did not have sufficient seats to reach down to Cat 4. Rhein-Main and Ramstein had not moved Cat 4 for over 30 days and there was no traffic into Lajes. I spent a lot of time on Autovon (now DSN), but things were tight everywhere going to the States. After a week, we needed to get back to the States, so we bought a commercial ticket into St. Louis.

We really enjoyed our stay at Mildenhall. Billeting was great at the Officers' Club. On about three occasions during the week, we had to check out in the morning and get back on the list. However, we always got back in. Food was good at the Galaxy (all ranks) Club; they serve three meals everyday. There is a lunch buffet at the Officers' Club, also good evening meals in the bar and on certain nights in the dining room. On July 4th, there was a picnic on base with food and cold drinks; also a fireworks display that night. One of the nice things about Space A is the people you meet. We met some interesting people as we all sat at the terminal, researching possible flights and sharing information about flights. After a week we felt we had known each other for a long time.

On July 8th, we took the free bus to Gatwick, flew TWA to St. Louis, found the car in good shape, and drove back to San Antonio.

This was our first overseas Space-A trip. Although we had to buy the ticket back to the States, we learned a lot and are planning more Space A travel in the future.

Sincerely,
Richard Kernaghan
LtCol, USAF, (Ret)

Dear Ann,

For three weeks my wife and I had no luck whatsoever in getting on either of the two flights per week out of Norton AFB. The C-141s were carrying hazardous cargo and no passengers were permitted. We called over to March AFB Pax Terminal and found they had two KC-135s going to Mildenhall/Saudi Arabia in two days with 5 seats on each. These were crewed by Air Force Reserves who were volunteering for extra active duty. They were from the 336th AREFS and called themselves RRATS for Riverside Refueling and Transport Squadron. I bet George Bush himself doesn't get such good service on Air Force One as we were accorded. We had a fun non-stop flight direct to Mildenhall.

The crew told us they were going onto Jedda but would be coming back through Mildenhall and would take us back with them if we were there. We were, they did and all of us got home to California on Thanksgiving Eve. Again, George Bush would have been jealous - we can't say enough about these people and we hope the good Lord keeps and protects them along with all our men and women in the armed forces.

Mildenhall was absolutely dead as far as passenger traffic is concerned. C-5s were leaving for Dover and other U.S. destinations with three or four people on them. The C-141 traffic to Charleston appeared a little tighter but still possible with a little patience. We have never seen that terminal so dead. Even Space-A government quarters were available, mostly during the week, but tight on week ends. We suggest if Mildenhall is not optimistic for that night try calling RAF Lakenheath. It is the home of the 48th Tactical Fighter Wing and you can reach their billeting office via DSN.

There is a government shuttle bus between bases (a 15 minute trip) and be caught at the Mildenhall AMC Terminal and you can be dropped off at the Lakenheath Billeting Office. The bus operates eight times each week day. The schedule is posted in the AMC Terminal. Weekends have fewer trips.

One item of interest to U.S. west coast Space-Aers is the weekly C-5 flight from Travis to Tinker to Dover to Mildenhall to Ramstein was supposed to have been reinstated about [10 November] but it had not happened nor did the Mildenhall AMC people have any update by [November 21] when we left. Hopeful passengers may want to call their applicable AMC desk to check.

Very Sincerely,
Dick and Betty Bunten
Alta Loma, CA.

EUROPE

Dear Ann,

For Catherine and me, the highlight of 1993 was our Space-A/EURAIL adventure, six weeks to eight countries, via Air Force cargo planes and some great trains!

We headed to Charleston (NC) where we set up our tent at the family campground. On the fourth day there we hopped a big C-141 to Lajes Field in the Azores. Category 4's - that's us - have an advantage going via the Azores as active duty persons who aren't stationed at Lajes cannot stop there unless they have passports (This increases the chances for catching a flight out of the Azores).

Terceira Island was a great place to be marooned four days while awaiting an opportunity to fly on to Europe. Although we were a planeload, everyone got quarters at $16 a night in wood-framed barracks which were very comfortable although not plush.

Four days after we arrived, a commercially chartered plane bound for Aviano AB stopped to refuel, and we left with it for our first visit to Northern Italy.

Aviano is at the foot of the Alps. The Mountain View Military Lodge is more than comfortable, and what a view of the mountain which rises above the base. We were able to shop in the exchange there, something we couldn't do on most of the bases in Germany.

European Trains: Mostly they're clean, very fast, and very modern; and there's one going where you want to go, when you want to go. But as with Space-A travel itself, country-hopping by rail isn't for sissies. It means toting your luggage up and down long flights of stairs, down from Track #15 and then back up to Track #3, for example, several times a day, and sometimes on the run.

From Aviano, we went east by train to Trieste, on the Italian Riviera, and then south to Florence to see Michelangelo's David before heading north through Switzerland.

Next we headed to Heidelberg where we stayed on base to rest a few days. We had excellent quarters in Patrick Henry Village. I had no problem using the AAFES there.

Most of our stay in Germany was in Southern Bavaria. From there we went east to Salzburg, Austria in order to catch the Orient Express to recross Germany westward to France's Alsace-Lorraine. Then, after a final swing through Switzerland we made our way back to Aviano AB for a flight next morning to the Azores. Four days later (we were on our way) back to Charleston.

Sincerely,
Joe W. McLaughlin

As Joe points out, Space-A and European trains often times are for the stout-hearted. PACK LIGHT.

Dear Ann,

We've recently returned from our first Space-A trip to Europe. We decided to be daring and also took our three children, ages 20 (a full time college student), 14, and 9.

We live a half hour away from McGuire AB, so we signed up for Space-A well in advance and revalidated our names on the list until the day we wanted to leave. We were fortunate to get a flight on the day we had chosen.

We flew on a DC charter to Frankfurt. The DC-8 is tight and crowded, but the price was right! We arrived in Frankfurt in the morning and took a train to Munich.

We chose to take our trip in late March. We found that the weather, although still winter, was generally mild and the crowds were thin. We don't think that we would want to travel Space-A or without reservations or itinerary in the high summer seasons with a party of five people.

We returned to Frankfurt upon the expiration of our train passes. Although we were fortunate enough to stay at the Rhein-Main AB hotel during our wait for a CONUS flight, Space-A is very limited. We would advise anyone who is waiting for a flight to be prepared to stay in a commercial hotel.

It seemed that many retirees waiting for a CONUS flight were distraught at the wait for a flight and the limited Space-A facilities at the base hotel. We were prepared for both, especially by reading your R&R reports and following your TML guide. Consequently, we had allowed a full week for the wait, and had enough money to sustain us through the wait. Because we were prepared, we weren't under the tension of an impending deadline of when we absolutely <u>had</u> to be back home. We knew what to expect so we relaxed and enjoyed it. Our nine-year-old son thoroughly enjoyed the video games at the airport, and the rest of us read or watched the planes from the Frankfurt am Main commercial airport.

It took us three days to get a flight home. We caught an unscheduled C-5 to Dover. We might have been able to get a DC-10 direct to McGuire the next day, but with a party of five, we thought it best to get seats when we had them. The C-5 was very roomy it was the first time I ever had so much leg room. The only minor discomfort was that our two children who sat next to the emergency exits were chilly. The blankets that were provided solved the problem.

Would we do it again? Tomorrow! The minor discomforts of Space-A travel are offset by the savings. We met other traveling retirees and learned that much of the success of the trip depends on being prepared for some minor inconveniences and not getting upset or inpatient. **Our trip would not have been possible at all without Space-A.**

An important note however, we were not prepared for the fact that retirees cannot use base PX or commissary facilities in Germany. We think that retirees traveling to Germany should know this in advance.

Patricia & 1SG William J. Jemison.

The Jemison's make several good suggestions; 1) As with the Boy Scouts - BE PREPARED, 2) Take extra cash for those unforeseen circumstances, and 3)Be flexible, it helps not to be on a tight schedule.

Editor's Note: *This next reader's Space-A trip was made a few years back, and the Athens Space-A facility mentioned (Hellenikon) is now closed. Colonel Sparks used the Athens facility as the "central point" for his trip. Now Adana Turkey could serve as a "central point" for most flights to Middle Eastern countries. Flights still leave Ramstein for Tel Aviv.*

Dear Ann,

In regard to some of the recent comments regarding Space-A travel to Southern Europe, Israel, and Africa (Egypt), I might offer the following in the hopes it could be of use to you in helping others.

I recently returned from a journey to Spain, Greece, Turkey, Israel, and Egypt, most of which was courtesy of Space-A. I flew out of Dover and got out of there within 24 hours of arrival. At Torrejon I found the people both at the terminal and at the guest house to be very friendly and helpful. I had no trouble getting quarters. Again, within 24 hours of signing up, I was on a flight to Athens. After touring Greece for two weeks I was able to get a flight in less than 24 hours to Izmir, Turkey.

I was unable to get a ferry from Turkey to Rhodes and Cyprus, so I got a Med-Evac flight back to Athens. I found Athens to be a good "central point" for making connections and the people there were quite helpful. After two days I was able to get a flight to Tel Aviv.

In Israel I joined a tour group, a plan I had before leaving home. It was well worth the cost because of the information the tour guide was able to impart regarding the whole of Israel and also since I was there over Christmas and New Years. Unable to get on an Air Sinai flight to Cairo, I travelled to there with two other Americans by taxi.

Nowhere have I seen anything listing AMC flights out of Egypt, but checking at the US Embassy I found a AMC flight desk and that there are weekly flights to Ramstein, Germany.

Not wishing to go as far north as Ramstein, I flew Air Ethiopia back to Athens. It appeared that there might be difficulty getting out of Athens this time so I took an unscheduled flight to Sigonella Naval Air Facility (Sicily). After being there two nights I got on a chartered DC-8 to Philadelphia, stopping at Naples and Rota.

All in all it was a wonderful trip and made possible and more pleasant by the US Military personnel, facilities, and equipment serving our country. For those who have the time and temperament, I recommend it highly.

Yours,
John M. Sparks
Col USAF (Ret)

Dear Ann,

Louise and I left Lighthouse Point, Florida in our old Chevy, [and] headed for Charleston, South Carolina. We had such an enjoyable trip [previously] that we thought we would try "Space-A" again. This time we planned to just fly around Europe at random, letting chance dictate our itinerary, and visit different countries as we came to them.

We tried to correct mistakes of our first trip, and to learn from the experiences of others. This time we carried umbrellas and raincoats. We packed only one duffle bag and two carry-on bags. We used small containers for shampoo, toothpaste and other necessities. Luggage can get burdensome in Frankfurt and Madrid.

As a precaution, we switched our records of traveler's checks. Louise carried mine and I carried hers. I also Xeroxed the first pages of our passports and both sides of our military I.D. cards. I had read about that somewhere and it just seemed like a good idea.

We arrived in Charleston on Saturday and stayed two nights at the Holiday Inn. I signed up for Turkey throughout the trip, reasoning that we could go as far east as possible without getting bagged for another paperwork fee. It just felt good to be in that terminal again and to get to know our fellow-travelers, most of whom were retired military.

Luck was not with us this trip insofar as getting a ride. We spent 10 days at Charleston, meeting all flights and scarcely any retirees were manifested. Luckily we were able to stay at base billeting for six of those nights, thus minimizing our expenses.

When it rains, it pours. On the tenth day, an unscheduled C-141 arrived with 84 seats and there were only 35 people boarding.

Since the flight terminated at Prestwick, Scotland, we all had to get off and make our own way from there. One couple rented a car and drove to Mildenhall. Another couple, from Louisiana, caught the train with us as far as Edinburgh. It was their first Space-A experience. They had just arrived in Charleston and one hour later their plane arrived. Talk about beginner's luck.

Louise and I toured Edinburgh and stayed at two excellent bed and breakfast establishments on Gilmore Street. Our Scottish hosts were friendly and hospitable.

Our three nights were memorable. I especially enjoyed the pub and the room across from it was brightly lighted and had no TV sets. In that situation, we found we were meeting people from all over Britain, as well as Australia, New Zealand, Canada and the United States. Staying at the Union Jack was our smartest move while we were in London.

Our dumbest move was getting on the underground (subway) during the rush hour. My wife had her purse picked between Embankment and Victoria Stations and lost money, traveler's checks, passport, military I.D., and Visa card.

Luckily we had all our numbers. It proved very helpful to have Xeroxed the passport and military I.D. By 1400 the next day we had canceled our Visa Card, received new traveler's checks (those were delivered to the hotel by courier) and were issued a new passport at the American Embassy. The embassy staff was unbelievably helpful, even calling Mildenhall and ascertaining that a new I.D. card could be made there. Incidentally, military I.D. authorizes one to eat at the embassy cafeteria. We had an excellent lunch there.

We stayed at the "Bird in Hand Motel" one night, and in base quarters the next, before boarding a Tower Air 747, which had 138 empty seats, for Rhein Main, Germany. Again, we cleaned out the terminal and had seats left over. I think a large part of my enjoyment of Space-A travel comes from the "rush" I get when they call my name at the counter and I realize I am about to actually board the aircraft. There is nothing like it!

From Rhein Main it is difficult to get anyplace without a car, but a sergeant, named Joe Ard, was kind enough to drive us all the way to Ramstein, his duty station. Beautiful scenery (especially of grape fields) and interesting conversation made the two hours go by very quickly.

That night we spent at the Pfaelzerhof Hotel at Ramstein. A Chinese restaurant provided a fine meal at relatively low cost.

The next day proved to be another disaster. I left the AMC terminal for a few minutes and missed the plane to Sigonella, Sicily. Then, somehow half a dozen other retirees and I got finessed into waiting all day long for base billets. At 1630 the boss of the place announced that there would be no rooms for retirees. I got the impression that those (foreign nationals) in charge had the idea that we (retirees) were trying to receive perks to which we were not entitled. At any rate, we wasted a day for lack of any system for allocating rooms in some logical way.

Finally, four of us hopped in a cab and stayed at the La Donola Motel just outside of Ramstein. That inn was quite modern and had a good restaurant. The other couple was from Oklahoma and had traveled throughout Germany. I think one of the greatest features of Space-A travel is comparing notes with fellow retirees. It is reminiscent of being on active duty.

Louise and I caught the plane for Pisa the next morning, along with a softball team stationed in Great Britain. The team was to play at Aviano, but Pisa was as far as the plane could fly. I think they finally boarded a train. Meanwhile, we decided to stay aboard our C-141 and go on to Torrejon, Spain. Since I had canceled our Visa card as a result of the robbery, we felt especially vulnerable insofar as having access to funds on demand. We began to realize our vacation would be shorter than expected.

Upon arriving at Torrejon, we knew the drill. We immediately went downtown for our "entradas" so we could travel on and off base. Then, we tried to find a room. The base billets were full, so we rode a cab all the way into Madrid, but not a single room was available. Apparently there were some conventions in town which, coupled with an increase in the number of tourists this year, made Madrid and the surrounding area very crowded indeed.

Spending all night in a passenger terminal gives one a chance to think. My main thought was that one night was enough. The next morning we caught the first flight to the States, a C-141 which stopped to refuel in the Azores.

Someday, I'd like to spend a few days just exploring the Azores island of Tercera, but we sensed a need to return home soon.

We arrived at McGuire AFB in the evening and caught a taxi-van to the Philadelphia Airport, where we rented a car for the drive to Charleston. Surprisingly, our car, which was parked across from the passenger terminal, was not at all damaged by hurricane "Hugo", and we were able to drive right off.

I must say that all-in-all this trip wasn't bad. Although we had a few hard knocks, we met a lot of people and saw some interesting places. the weather couldn't have been better. Also, there is something about being on those Air Forces Bases again that really takes me back to other times, to another life, to memories I don't want to fade altogether. I can't wait to do it again.

James P.Meagher
Lighthouse Point, FL

James gives excellent advice when he says to make copies of both your passport and military ID card.

GERMANY

Dear Ann,

My wife and I took our first Space-A trip since I retired 8 years ago. We traveled by SAC Tankers from Grand Forks AFB to Mildenhall RAF, England. We got a C-141 out to Ramstein the next morning and shuttle bussed over to Rhein Main where our son is teaching school (DOD) for two years. Five weeks later we headed for Grand Forks (and our automobile.) Just after July 4th we found openings for CAT 4's very rare but we did get to Mildenhall after being turned down only twice. We waited five days at Mildenhall for a KC-135 ride. Thanks to the Plane Commander who opened up (space) to take 40 passengers we made it. We would think twice before using Space-A again in midsummer.

We enjoyed the lodging and dining facilities at Mildenhall although I was up at 0400 every morning to be early on the 1700 Billeting list. We never had to go off base.

I have one gripe. In reading many R&R's I had never noted it pointed out that commissary and BX privileges were not available to retired service people. I believe this should be pointed out to Space-A travelers, not to discourage them but to avoid unpleasant surprises overseas. As I renew my subscription I want to thank you however for your informative and interesting bulletin.

Sincerely,
Lester Swanson
LCDR (USNR)

*Two interesting points here, again the extra mile that AMC personnel go, and the tip to try and fly in the off-peak season. We would also like to point out that the priveleges of active and retired personnel is covered in Status of Forces Treaties with each country. We have published this information in our U.S. Forces Travel Guide Europe, and Near East Areas and more recently in our **R&R Space-A Report®**.*

Dear Ann,

We just returned from three weeks in Europe. Although we have visited there often, this was our first Space-A effort. Thanks to your fine publications and newsletters, we were extremely successful and had a wonderful time.

We drove from Virginia Beach to Dover and back about three weeks before we were ready to depart. We wanted to go to Germany and Dover seemed to be the best place to leave from. Registration was easy. No forms to fill out at all. Just show your ID card and they put you in the computer and give you a copy of the registration showing Julian date etc. We could have left that day if we wanted.

When we finally decided to leave we showed up at Dover "Flight Ready" at about 0800 which is way too early since most flights to Germany leave late at night. There were people who had waited up to four days there. This was primarily due to Hurricane Disaster relief missions taking priority (Florida and Hawaii). Dover has great facilities though and the waiting isn't bad. We were lucky and got a flight on a C5A Galaxy to Ramstein AFB at 2200 that night. Parked our car in Dover's long term parking lot. Your description of the ladder on the C5 is accurate but it wasn't a bad climb. The plane was full and offered one hot meal which was Steak! We arrived in Ramstein at 1210 the next day just in time to miss the 1200 shuttle bus to Rhein-Main where our rental car was located. Registered for a return trip at Ramstein then finally got to Rhein-Main at about 2000 that night. Registered there for a return trip too. Ramstein had a form and a date stamp machine. Rhein-Main has a computer setup similar to Dover.

We spent three weeks, mostly in Austria although we did visit Heidelberg and Regensburg Germany. For our return trip we got into Rhein-Main and turned our car in about 1600. We waited nearly two full days because we were trying to get to Dover/Philadelphia and turned down other flights. As it turned out, we could have taken the others and it would have worked out. Everyone at Rhein-Main is friendly and helpful. The Hotel on the base generally has room although we did not use them since we wanted to be there for unscheduled flights. Finally got a C141 scheduled for Azores (RON) then Dover. Many people didn't take this flight since they did not want to RON in Lajes, Azores. That's a mistake. We got on the plane which had a small amount of cargo and about 150 seats. There were only 14 of us passengers aboard and it was a great flight. When we got to Lajes, I asked what time the next day we would be leaving and was informed that we would not RON but would leave in about an hour and a half after refueling. Just time for a great lunch and some souvenir shopping. We got back into Dover about 1900 and found our car safe and sound.

To summarize, thanks to your great material, our trip was successful and we'll definitely try it again. We have recommended you to a number of people and given them referral slips. I hope they follow through.

Thanks
Ed DeLong

Ed, like many of our other readers, notes that the Azores is a great place for Category 4's to catch a flight.

HAWAII

Dear Ann and Roy:

After a wonderful eight days in Oregon we headed down the coast for San Francisco. As a lark we decided to see what was going on at Travis. We detoured 30 or so miles to check out the action at the AMC and try to get a room for the night. We arrived around 1400 and checked the Space-A book to be sure our names were entered into the computer (they were). On a whim we asked the desk what was available that afternoon. The Air Force sergeant said "We have a C-5 flight leaving for Hickam AFB in Hawaii and if you can park your car and get your luggage into the check-in area in 30 minutes there are seats available." (Two earlier flights that day had taken every one that was waiting).

We were caught completely off guard. We had planned to go back to Pismo Beach, CA. the next day and end our eight day vacation but here was a wonderful opportunity. Talk about impulsive, to make a short story shorter we decided to get aboard. There were 42 passengers on the 73 seat flight so many of the families let their kids lay down and sleep on the flight. As usual there was a hot meal served and it cost us only $2.30 each for a TV dinner.

We landed at Hickam AFB and phoned the Pacific Marina for a room and pickup service. Billeting was full. The next day we got a hotel room and rented a 1994 Mustang. The car was a honey and the room was about two blocks from Waikiki Beach. For eight days we did the usual tourist things (i.e. visited the USS ARIZONA MEMORIAL and cried for the lost young men and women during the excellent movie that showed the events leading up to the Dec. 7, 1941 bombing).

At the end of seven days we decided to try and catch a flight back to Travis. (Your publication has warned us not to travel during the summer on a Space-A basis and we now know what you meant. Not only was school just out but there were lots of people with change of duty station orders and plenty of Active Duty people just touring Hawaii. Last but not least was the 25,000 men and woman who were out in the Pacific near the Hawaiian Islands because of the RIMPAC exercise that included troops from five countries. When we walked into the Hickam terminal there were about 130 people waiting for flights with many of the active duty personnel going to Travis. When the first show time was called for 31 seats on a C-5B, people started arriving from all over the place and it became obvious that it would be a long time before they even got to the Category 2B people much less the Cat 4 group. (There were probably 35 retired people waiting with some of them telling me they had been waiting for 5 days and it looked to them as if the lines were getting longer rather than shorter. More Active Duty and Dependents were coming in for flights as other flights were cancelled.) My wife and I could see that it was going to be a problem getting out so we caught a commercial flight to California.

At the airport baggage level we found the Sacramento bus and in one hour and 30 minutes it stopped at Travis AFB where we picked up our car. The last bus leaves at 11:15 p.m. A very happy ending to an impulsive, unexpected and wonderful holiday. It's probably good for 64-year-old grandparents, and retirees to do some things on the spur of the moment, once in a while.

Best wishes to all of you Space A readers and good luck on your next trip.

Sincerely,
Don and Jean

ISRAEL

As you will see from these letters, the YMCA in Jerusalem is a great place to stay.

I have made several trips to see friends who retired in Israel. My route - Dover AFB (never waited more than two days) to Torrejon AB Spain. Again, two days is the limit I spent waiting there. From Torrejon I flew to Athens (Hellenikon AB); spent three days at a recommended hotel and flew out to Tel Aviv.

There is a detachment or Air Attache in Tel Aviv. If you go to Jerusalem make reservations at the YMCA (across from the posh King David Hotel). It is a popular place - very imposing on the outside, very clean, plain rooms and good food.

Also keep your bags with you (even if going to (the) telephone, getting change or a cup of coffee) while in the airports. The security guards are very strict and investigate and check any suitcases or packages left unattended.

Sincerely,
Jeanne C. Stevens

Dear Ann and Roy,

[I] had inquired several times when passing through Athens about AMC flights to Israel and was informed that there were generally two scheduled flights a month, although special unscheduled flights were rather more frequent.

Because we had been warned that hotel space in Jerusalem was at a premium, we decided to reserve accommodations at the Aelia Capitolina (YMCA) Hotel over a month in advance. In actual fact this is a three star hotel with swimming pool and (is) a short, four block walk from the Damascus Gate of the Old City.

A virtually unparalleled grid lock faced Category 4 travelers at Dover, with some old timers reported as having been waiting three weeks with no Space-A movement whatever. We ended up flying *Swissair* to Zurich, then by train to the AMC base in Aviano. From there we flew to Sardinia, Naples and Athens.

After landing in Athens we took a ferry to Haifa, Israel. We arrived in Haifa almost exactly two days after departure. Buses left regularly for Jerusalem; the fare was cheap. A taxi took us from the terminal to the hotel. We spent seven days wandering through the four quarters of the Old City.

Departing from Tel Aviv we went to Ben Gurion Airport and got a flight via AMC on a C-141. They got about 10 of us out on a flight the authorities had just about decided to close to passengers.

Sincerely,
Blair Walliser
CAPT, USCG (Ret)

Dear Ann,

A dream my wife and I always had was to go to the Holy Lands and see all the places Jesus walked, talked, where he was born, died, and preformed miracles. This would really be great!

I decided to go in February because flights are not crowded and it would not be too hot in the Holy Lands at that time of the year. We had no trouble getting a flight out to Rhein-Main From Charleston. We went by the shuttle bus to Ramstein and checked into base billets just a block away from the terminal.

We got a flight to Adana, Turkey on a C-5A, but could not get base quarters because of an exercise on the base. We caught a contract flight to Athens and stayed at the military hotel by the beach.

On Wednesday we got the C-130 to Tel Aviv, Israel. Customs was no problem, but they told us not to leave our bags unattended or they would be destroyed. Outside the air terminal are buses and limos that will take you to Jerusalem. We stayed at the YMCA which is within walking distance to the Old City.

We booked our tours right at the desk for the Dead Sea, Masada, the Bedouin Market, Nazareth, Capernaum, Bethlehem, and the Sea of Galilee. All were cheap and wonderful to see.

After spending five days in Israel, we got a flight back to Stuttgart, Germany on a Lear Jet. From Rhein-Main we went to Lajes on a C-141 back to Charleston.

MSgt Frank J. Terkovich
USAF (Ret)

SCANDINAVIA

I went to McChord AFB on the morning of July 14th, expecting to get on a plane bound for Germany. No such luck. The plane normally scheduled for Travis-Tinker-Dover-Ramstein was diverted to Hill AFB, Utah. I didn't want my trip to end up in the middle of the desert, so I came home.

The next Friday I went back to McChord and was determined to go SOMEWHERE. I left McChord at 11:40 a.m. on a C-141 for Tinker AFB, Oklahoma. It was bound for Dover, Delaware and Germany, but didn't have any room for standbys. At 7:30 p.m. (PST) took off for Andrews AFB, Washington, DC, aboard a Navy C9 (DC-9) that originated at Oak Harbor and came in by way of Miramar, California. The C-141 is a cargo plane and was so noisy that they issued us earplugs. The "Spirit of Oak Harbor" (C9) is a regular civilian-type transport plane. Much nicer.

Arrived at Andrews about 11:00 p.m. (Washington, DC time). The passenger terminal at Andrews was closed for the night, and as it was Reserve Training Weekend, there weren't any billets at the BOQ (Bachelor Officer's Quarters). The car rentals agencies were all shut down for the night, so four of us rented a taxi to take us to the AMC terminal at Dover, Delaware. We arrived in Dover about 3:00 a.m., had breakfast, and tried to sleep in the most uncomfortable chairs ever made.

[I] did not get out Saturday or Sunday. The next scheduled flight was 0320 Monday. After 73 seats were filled and about 20 Category 2 people were left, it was apparent that Category 4 people weren't going to get off. [I] decided to take a commercial flight from Baltimore to Frankfurt. 1LT Luke Fuka, who was on the same flight, rented a car to drive to Langley AFB, Virginia and dropped me off in Baltimore. (I) boarded the Airbus A320 and landed at 7:00 a.m. (Tuesday) in Frankfurt. The (AMC) plane for Oslo was scheduled to depart at 9:30 a.m. Wednesday.

I anticipated no problem getting on the Oslo plane, as they had 14 seats available. When it came time to board however, I was the 15th person trying to board. The clerk called out to the plane and they found an extra seat for me. In the air at 10:00 a.m., about two and a half hour trip aboard a C-130 Globemaster. I arrived in Oslo about 12:40 p.m. at Fornbu Airport on the side of the field shared by the U.S. and Norwegian Air Forces. A $5.00 taxi fare got me to the airline terminal and I bought a ticket for Trondheim on Braathens Airways for 860K ($123 on VISA) to depart at 2:05 p.m. Forty-five minutes later I am in Trondheim and set to begin my vacation with my Swedish cousins.

[On the return trip I caught a] General's plane...a C-21 Gruman Gulfstream...going straight to Washington, DC, with a fueling stop in Gander, Newfoundland. It was carrying a Colonel, a crew of five and an Air Force stewardess, and the only other passengers were an Air Force captain and his wife and myself. The trip took nine and a half hours, including the fuel stop. We landed at Andrews AFB, Washington, DC (on Thursday) at 3:15 p.m. EST.

There weren't any planes out Thursday night. [I met a retired RADMU who] made a couple of phone calls. He found a Navy Aero Commander heading for New Orleans. We high-tailed it to the Navy side of Andrews Air Base. [The RADMU's] phone calls had gotten us manifested on the Aero Commander to New Orleans leaving that afternoon, on a Navy C-9 leaving New Orleans on Friday night for North Island, and on another C-9 leaving North Island Saturday morning for McChord AFB.

We arrived in New Orleans on schedule in the (eight passenger) Aero Commander and checked into the BOQ. The next (day)...we boarded the C-9 on schedule at 2230 and flew to North Island landing about 3:00 a.m. Saturday. After we had been in the terminal about an hour, the cafeteria opened for breakfast. The plane was scheduled to leave about 9:00 a.m. with a stop in Alameda. Upon arrival (at McChord AFB) I retrieved my truck and was home by half-past two.

Carl J. Nordstrom III

SPAIN

Dear Ann,

We made showtime at Dover within a half hour of our arrival. The passenger agents at Dover stamped an "Entrada" in our passports, and they assured us this was all we needed. We flew to Torrejon in a C-5 with half the seats empty. On arrival at Torrejon I showed our "Entradas" to the Spanish Customs agent; he insisted Dover had no authority to grant Entradas; and he stamped his Entrada in our passports.

There being no scheduled plane to Rota for five days we took the free bus to Borrajas Airport from which we taxied to the rail station. There were no seats left on the train! We trained back to Torrejon [136 pesetas each; 1 1/2 hours]. At the gate to the air base a very friendly English speaking Spanish guard looked at our ID's and assured us Entradas were no longer necessary! In the 30 days between our trips the car rental agency at Torrejon Base had reopened; there was a bus to Borrajos Airport; and the Spanish people to whom we spoke were much friendlier. I don't know if the definite date for the closing of the American part of the air base has removed an irritation or the coming of the 1992 Olympics is the answer, but for the first time we were made to feel like welcome guests in the Madrid area.

We got a ride on an unscheduled plane to Rota and tried to complete the MWR tours we had come back for. A pouring rain cancelled out the trip to the Gypsy flea market. A strike at the Sherry bodegas cancelled out that trip. We rented a car and saw the show at the Andalusian Riding academy in Jerez.

We visited Seville but didn't allow enough time to see much, and the street names on our map did not match those on the ground. The Tangiers trip was still on, and we signed up for that one. There were two full busses out of Rota to the ferry where we turned our passports over to the customs agent on the ferry. (In case anyone is curious there is no problem entering Morocco with an Israeli stamp in a passport). Two busses were waiting for us on the Tangiers side each with a local guide. Ours spoke excellent English, but the PA system wasn't functioning. We were warned repeatedly to keep together. The lunch and entertainment were excellent. When we got to the shopping part of the trip we were taken to a rug shop where we remained for over a hour. When we left we were told we were late, and we were rushed through the streets to our busses. We arrived at the ferry about fifteen minutes late. We were told the Captain was angry and to board quickly. Our passports were returned. Those who knew asked for their passports to be stamped when they handed them in. Those who didn't ask (including us) got their passports returned unstamped. We returned to Rota on schedule.

Two days later we got seats on a C-141 for Norfolk and McGuire and home, again.

Best regards,
Stanley Goldberger
Col AUS (Ret)

FRIENDLY PEOPLE & A TIP

To the Crawfords,

Darrell and I wanted to share our experiences with you on our recent trip to Guam, Singapore and Yokota, Japan.

We feel flying Space Available has been a delightful experience. We have met such lovely people. Three different couples went out of their way to help make our travels more comfortable. In Japan we stayed in the home of a retired couple, who worked in Japan, for the night. There weren't quarters on base and no taxis available.

In Guam (our daughter lives there) we caught a C-5 for Singapore. It was going there for the Air Show in Singapore. It was a spur of the moment decision on our part. We didn't have hotel reservations in Singapore. Aboard the C-5 one of the officers asked if we had accommodations. We told him no. He said it would be difficult to get them with everything happening in Singapore - Air Show, Chinese New Year celebration etc. But the military would try to get rooms for us. They assigned a military man in Singapore to us and he helped us go through the airport and onto a taxi. We had just perfect hotel accommodations at the Carlton Hotel for 2 days. Then we had to transfer to the Van Coolla which was very old but clean. I have enclosed the Agency's card. They help the military personnel with hotel etc.

Anyway, we have only positive things to say about the military personnel - very friendly, concerned and informed - a wonderful time for us.

We hope to use your Military Living Temporary Military Lodging Around the World to continue on our Space-A travels.

As ever,
Colonel and Mrs. Gallear

PS. It pays to stay around or check in at the counter several times a day even though there is nothing scheduled. We caught an unannounced contract 747 from Hickam to Guam with the next scheduled plane three days away. The C-5 from Guam to Singapore was unannounced until the morning it left (we got on only because it had a breakdown and couldn't leave on time). We had planned on going to Yokota and then to Singapore; the straight shot was much better.

SPACE - A + TIME & PATIENCE = SUCCESS

Space-A remains tops if one has the time, patience and a reserve of cash to wait for this type of transportation. The personnel are among the finest and should be commended for a job WELL DONE.

A Concerned Reader

SPACE - A LESSONS LEARNED

Dear Ann & Roy,

Just returned from a five week Space-A European trip. We left Travis February 7th on a C-5 and got the last two seats. Only five Category 4's on board. The 73 temporary seats posted on the monitor was instantly reduced to 36 at Showtime due to the women's Air Force Basketball teams returning to their various home bases after a tournament at Travis. We "made" Ramstein on schedule and traveled by train to Prague and Budapest.

We then went to Mildenhall to visit our daughter and had no luck catching the one flight a week back to Travis on Sundays. Too many active duties. Went back to Ramstein and bussed back and forth to Rhein Main for next available seats to CONUS. What we ran into was NO passenger movement due to freight commitment and a sign behind the AMC desk showed no Category 4 movement to CONUS since February 26th and this was March 7th.

An item to note: Watch out for Cat 2's because they travel Space-A right after pay day. On March 8th we caught a B747 Northwest Airline charter from Rhein Main to St. Louis. There were about 15 Cat 4's aboard and the plane was chock full. My wife and I then [took a shuttle bus and] got beautiful quarters at Scott AFB, 45 minutes away.

I put [my wife] on a commercial flight to Ontario, CA the next day and I went back to Scott to wait out a Space-A MEDEVAC to Travis but too many uniforms. Finally, on March 12th [I] caught a Federal Express charter from St. Louis to Oakland, CA which was going on to Japan and Korea. Plenty of seats for all comers to Oakland. (I] got to Travis at 0100, March 13th where my car was, stayed over night at a commercial motel and drove the 400 miles down home which is within shouting distance from Norton.

LESSONS LEARNED

Lesson 1: Start travel to Europe in late January, not early February.

Lesson 2: Watch out for paydays.

Lesson 3: Scott has excellent quarters and they treat you like a king but Space-A opportunities for Cat 4's is limited west-bound. It is good for middle of the country destinations and easterly to McGuire and Andrews.

Sincerely,

Richard C. Buntenm
CAPT USN (Ret)

LOST LUGGAGE

Dear Ann,

On my most recent Space-A trips I've had the two most exciting experiences of "lost baggage"!!

"Baggage loss number one" was a result of our being "bumped" from a C-5 flight out of Kadena through Tokyo to Travis. During an 18 hour overnight crew rest at Yokota all passengers were bumped off the flight and the mission changed. Since we were off base at Tarra Hills Lodging, we did not claim our luggage. Later in the afternoon our bags had disappeared, but nobody knew where to! (our bags were off loaded and left in the lock-up AMC baggage section)

From this point on it was a hilarious guessing game as to where our bags had been taken. We were assured that they had been flown to Guam, but later found out that they had been on-loaded again to the same C-5 (whose mission had been changed) and flown to McCord AFB. My wife and I caught a later flight to Travis and filled out the enclosed "Baggage Irregularity Report." The bags did arrive with all contents intact, but we were not contacted by the AMC personnel at Travis.

Just three weeks ago my son and I took a Beale KC-135 flight to Mildenhall, and on arrival one bag (an Air Force blue colored duffel bag) was not off-loaded from the plane. Two days of searching and no bag yet!! After two more days the bag finally appeared - it had been mixed up and delivered with crew baggage!

What are the morals of these stories? First, if at all possible hand-carry baggage. When "bumped" off a flight, claim your off-loaded baggage without too much delay, from the AMC locked baggage section. Try to avoid bags which have "look-a-like" appearance with crew luggage!!

Kindest regards,
Robert N. Class

THE GOOD,

Dear Ann,

We were able to get a KC-135 refueling tanker from Seymour Johnson AFB in Goldsboro, NC., non-stop to Fairbanks Alaska. There were 55 seats available and all 55 were taken with some people left over. Accommodations were a bit austere, but the crew was very courteous and accommodating. We were more than pleased with the flight.

Accommodations at Eielson AFB at Fairbanks were excellent. The billeting personnel were most courteous, and took care of us whenever possible, which was for most of our stay there.

The KC-135 on which we flew up was not returning for a month, and we didn't want to stay quite that long. We learned of a flight to Macon Georgia. This we signed up for as it would get us fairly close to our starting point. We learned [later] that it would be delayed for about three days. However they offered us an Air National Guard flight to Bangor Maine. Another KC-135 tanker. We were successful getting on that flight. Again the crew was very friendly, and allowed us to observe the refueling of a plane from Guam through the boom operator's window at the rear of the plane. We had to stop over at Elmendorf AFB to refuel and spend the night. Again the billeting personnel were courteous, and gave us excellent accommodations.

When we arrived in Bangor, I sent my wife home commercial, and bussed to Brunswick NAS where I was able to pick up a Navy reserve plane to Norfolk. From there I bussed down to Seymour Johnson AFB and picked up my car. Accommodations at both Naval bases were excellent.

Needless to say we are tickled to death with our first experience, and can't wait to try again.

Sincerely,
Joseph W. Westbrook, Jr.
LtCol USAF (Ret)

THE BAD, AND THE MALODOROUS (SMELLY)

Dear Ann,

I (am writing) to you about the troublesome subject of psychopathic medically retired Space-A travelers. Well this problem has affected me personally. I'm finishing a month long trip to Australia and on the very first leg of the trip, March AFB-Travis-Hickam I was subjected to body odor stench from a 400 lb Space-A passenger who took up two seats. I'm surprised that the Air Force general and his family who were riding along didn't protest. Ann, this guy stunk up the whole KC-10 passenger area. When we got to Hickam I complained to the AMC passenger counter supervisor. They spoke to the man who showered and changed his fuming clothing.

Another incident was when we were at Richmond, Australia. All active duty and retired Space-A pax were issued tickets, cleared customs, etc. Later all retirees' tickets were torn up, but the two active duty boys got to fly back to Travis. We were told that the pilot refused to take any retirees. Your guess is as good as mine of what caused the pilot to keep retirees off his plane and fly home with all those empty seats.

Also regarding this subject, I recently started work as a volunteer at a USO. During my initial interview I asked what was the biggest problem that I might encounter. The unexpected response was "retirees".

The Space-A benefit isn't chiseled in stone. We could easily lose it with a few retirees who suffer from mental illness/behavioral problems ruining it for all of us.

Maybe some of us are rude, rowdy or create a mess. We can all try harder to be polite to and thank the air crew for the ride.

R. D. Graham

Please note that all opinions expressed in the letters in the Reader Trip Reports section are those of the letter writers and not of the publishers.

A tanker refuels a fighter aircraft. Space-A passengers tell R&R HQ that tankers are great for Space-A travel.

APPENDIX A: SPACE-A AIR TRAVEL REMOTE AND ONE TIME SIGN-UP

REMOTE & ONE TIME SIGN-UP

The following is a list of Air Mobility Command Terminals/Stations Providing Space-A Travel and their FAX numbers for Remote Space-A Travel Sign-up, effective 1 July 1994.

Terminal/Stations	Telefax (FAX) Number
Andersen AFB, Guam	(671) 366-2079
Eielson AFB, AK	(907) 377-2287
Elmendorf AFB, AK	(907) 552-3996
Hickam AFB, HI	(808) 449-8108
Kadena AB, JA (Japan)	011-81-611-734-3048
Los Angeles IAP, CA	(310) 363-2790
McChord AFB, WA	(206) 984-3110
Osan ASB, RK (Korea)	011-82-333-661-4897
Scott AFB, IL	(618) 256-3066
St Louis IAP, MO	(314) 263-6247
Travis AFB, CA	(707) 424-2048
Yokota AB, JA (Japan)	(DSN)(315) 225-9768 011-81-425-52-2511-EX-59768
Andrews AFB, MD	(301) 981-4241
Dover AFB, DE	(302) 677-2953
Charleston AFB, SC	(803) 673-3060
Charleston IAP, SC	(803) 566-3845
McGuire AFB, NJ	(609) 724-4621
Philadelphia IAP, PA	(215) 897-5627
Grand Forks AFB, ND	(701) 747-6538
Plattsburg AFB, NY	(518) 565-6577
Norfolk NAS, VA	(804) 445-6578
Lajes Field (Azores), PO (Portugal)	011-351-95-52101-EX-4255 ***(Located in Command Post)***
Rhein-Main AB, GE (Germany)	011-49-69-699-6309
Ramstein AB, GE (Germany)	011-49-6371-47-2364
Sigonella NAS, IT (Italy)	011-39-95-86-5211
Aviano AB, IT (Italy)	011-39-434-66-7782
Howard AFB, PN (Panama)	011-507-8-43848
RAF Mildenhall, UK (United Kingdom)	011-44-638-54-2250
Incirlik Airport (Adana), TU (Turkey)	011-90-322-316-3654

PROCEDURES FOR REMOTE SPACE-A TRAVEL SIGN-UP (USE THE REMOTE SIGN-UP FORM FOR APPLICATION)

The Assistant Secretary of Defense (OSDUSD-TP), gave approval to USCINCTRANS on 30 March 1994 to implement remote sign-up for space available (Space-A) travel and sought service headquarters planned implementation instructions/guidelines. **The following (summarized) procedures for this imitative are effective 1 July 1994.**

1. ACTIVE DUTY MEMBERS OF THE SEVEN UNIFORMED SERVICES:

A. Telefax (FAX) a copy of the applicable service leave (or pass) form (**AF Form 988, DA Form 31, NAVCOMP 3065 and NAVMC 3 and others service leave (or pass) forms from USCG, USPH and NOAA).**

B. A statement that required border clearance documents are current, i. e. I.D. cards for sponsor and eligible dependents (family members); passports for sponsor (if required) and dependents (family members); Visas for sponsor and dependents (family members) and immunizations (PHS-731, I. International Certificates of Vaccination and II. Personal Health History), as required for all travelers.

C. A list of five desired country destinations (5th may be "ALL" to take advantage of opportune airlift).

D. The telefax (FAX) should be sent on the effective date of leave (or pass): Therefore, the telefax (FAX) header will establish the basis for date/time of sign-up.

E. Members will remain on the Space-A travel register for a period of 45 days or upon expiration of leave (or pass), whichever is sooner. As an option the Services (USA, USN, USMC, USCG, USAF, USPH, NOAA) may designate a central point of contact to assist members by answering basic questions and ensuring information is correct to minimize delayed sign-up. (Note, none have been so designated).

F. Mail (United States Postal Service) and Courier (Base/Installation official distribution) entries will be permitted. The Air Mobility Command (AMC) has indicated that commercial courier/delivery services such as United Parcels Service (UPS) Federal Express (FEDEX) are not acceptable media for filing your application for Space-A air travel. Upon receipt, the service leave (or pass) form can then be stamped with the current date/time (please keep in mind that mail on Military Bases goes through distribution channels and may take longer than normal mail). Note: Active Duty Members on pass my utilize this enhancement. Telefax (FAX) a request indicating desired destination, Name, Rank and inclusive dates of pass.

2. ACTIVE STATUS MEMBERS OF THE RESERVE COMPONENTS:

A. Telefax (FAX) a current copy of their DD Form 1853, **"Authentication of Reserve Status For Travel Eligibility".**

B. A statement that border clearance documents are current, if applicable for United States Possessions.

C. A list of five desired destinations (No foreign country destinations; 5th destination may be "ALL" to take advantage of opportune airlift).

D. Active Status Members will remain on the Space-A register for a period of 45 days. Note: Active Status Members of the Reserve Components may only register for travel to/from the CONUS and Alaska, Hawaii, Puerto Rico, The U. S. Virgin Islands, American Samoa and Guam. **Dependents (family members) of Active Status Reservist do not have a space available travel eligibility.**

E. Mail and Courier entries will be permitted. Upon receipt the DD Form 1853 can then be stamped with the current date/time (please keep in mind, mail on military bases goes through distribution channels and may take longer than normal mail).

3. ELIGIBLE RETIRED MEMBERS OF THE (SEVEN) UNIFORMED SERVICES:

A. Telefax (FAX) a request to the desired aerial port(s) (station) of departure giving five desired destinations (5th destination may be "ALL" to take advantage of opportune airlift). The telefax (FAX) data/time header will be the basis for the date/time of Space-A travel sign-up.

B. Retirees may remain on the Space-A travel register for a period of 45 days.

C. Mail/Courier entries will be permitted. Upon receipt the request can be stamped with the current date/time (please keep in mind that mail on military bases goes through distribution channels and may take longer than normal mail).

4. MEMBERS OF THE RESERVE COMPONENTS (GREY AREA RETIREES) WHO HAVE RECEIVED NOTIFICATION OF RETIREMENT ELIGIBILITY BUT HAVE NOT YET REACHED AGE 60:

A. These Members are limited to the same travel destinations of Active Status Reserve Members. **Dependents (family members) of these Reservist do not have a space available travel eligibility.**

5. On page 77 is a list of AMC Terminals (Stations) having the best capability of providing Space-A travel. Units listed in USAF, AMCP 76-4 will also provide remote Space-A travel sign-up. The Services (other than USAF/AMC, USA, USN, USMC, USCG and other USAF) are requested to augment this listing with their Base/Installation manifesting agencies capable of providing this service. **Note: Remote Space-A travel sign-up is the sole responsibility of each member unless a (Uniformed) Service designates a single Point of Contact (POC) for a specific installation. Note: Uniformed Services have not designated single POC's.**

6. Please note that **worldwide Space-A sign-up in person and at self-service counters remains available at Services terminals and stations which operate Space-A registers.** Times available for registration in person are based on local operating hours.

ONE-TIME SIGN-UP FOR SPACE AVAILABLE PASSENGERS

Passengers traveling Space-A on military and contract charter aircraft **now (since late May 1994) can retain their initial date/time (Julian date) of sign-up when traveling through more than one destination/station (traveling in the same general direction, i.e. East to West, North to South) to reach their final destination.** The new procedure is the result of a recommendation made by the U.S. Air Force's, Air Mobility Command (AMC) and approved by the Assistant Under Secretary of Defense.

In the past, travelers received a new sign-up date at each stop on their way to their final destination, which caused some to say that those stationed or living at or in the vicinity of the en route location had an unfair advantage. **Under the new rules, passengers still are required to sign up at all en route stops, but they keep their date and time of sign-up from their originating location.** For example a passenger who originates his/her travel at Incirlik AB (ADA), TU to Rhein-Main AB (FRF), GE, and plans to continue straight through to CONUS, will get priority over people starting their flights at Rhein-Main AB, GE.

The process is not automatic. **Passengers must still sign-up at all stops to continue their Space-A flights and retain their original sign-up date.** However, passengers receive an "in-transit" stamp on their travel order or boarding pass indicating date, time and location where they entered the system. This stamp identifies en route travelers to terminal personnel and gives the travelers priority on subsequent flights.

There are restrictions. **Passengers traveling (hopping) by Space-A Air through terminals/bases for extended visits will loose their transient status.** For example, if the above passenger gets to Rhein-Main AB, GE and takes six days of leave/pass/vacation there, he/she will get a new date and time for his next Space-A flight.

APPENDIX B: INTERNATIONAL CIVIL AVIATION ORGANIZATION (ICAO) LOCATION IDENTIFIERS AND FEDERAL AVIATION ADMINISTRATION (FAA) LOCATION IDENTIFIERS (LI) CONVERSION TABLES

The location identifiers (LI's) used in this book are the Federal Aviation Administration coordinated three letter LI's for the United States, its possessions, and Canada. Foreign country LI's have been coordinated by the Department of Defense. An LI represents the name/location of an airport/airbase. They are considered permanent (changes are made for air safety only) and cannot be transferred. The original LI remains in effect even if it becomes necessary to change the name of a given facility.

The International Civil Aviation Organization (ICAO) has established an international location indicator which is a four letter code used in international telecommunications. If the ICAO is shown on the departure board or other displays, look under the incode for the ICAO listed alphabetically to find the three letter Location Identifier (LI), the local standard time (LST), and the clear text name and location of the airport. If the LI is shown on the departure board or other display, look under the decode for the LI listed alphabetically to find the four letter ICAO, LST, and the clear text name and location of the airport.

The ICAO/LI's listed below are primarily used to identify military stations/locations around the world. Some of the stations listed may not be active at all times, and all worldwide stations may not be listed. Local standard time (LST) gives the difference in LST from Greenwich Mean Time (GMT).

INCODE (ICAO vs LI)

ICAO	LI	LST	STATION/NAME	LOCATION
AAAD	ADL	+09:30	ADELAIDE APT	WEST TORRENS, AU
ABAS	ASP	+09:30	ALICE SPRINGS RAAFB	ALICE SPRINGS, AU
AGGH	HIR	+11:00	HENDERSON IAP	HONIARA,SI
APLM	LEA	+08:00	*LEARMONTH RAAFB	EXMOUTH GULF, AU
APWR	UMR	+09:30	WOOMERA AIR STATION	WOOMERA. AU
ASRI	RCM	+10:00	RICHMOND RAAFB	RICHMOND, AU
ASSY	SYD	+10:00	SYDNEY IAP	SYDNEY, AU
BGSF	SFJ	-03:00	*SONDRESTROM AB	SONDRESTROM, GL (DN)
BGTL	THU	-04:00	THULE AB	THULE, GL (DN)
BIKF	KEF	+00:00	KEFLAVIK APT	KEFLAVIK, IC
CYAW	YAW	-04.:00	HALIFAX SHEARWATER	HALIFAX (NS), CN
CYQX	YQX	-03:30	GANDER IAP	NEWFOUNDLAND, CN
CYYR	YYR	-04:00	*GOOSE BAY AB	GOOSE BAY NFLD, CN
CYYT	YYT	-03:30	STJOHNS APT	STJOHNS NFLD, CN
DRRN	NIM	+01:00	NAMEY IAP	NIAMEY, NG
EBBR	BRU	+01:00	BRUSSELS NATL	BRUSSELS, BE
EDAF	FRF	+01:00	RHEIN-MAIN AB	FRANKFURT, GE
EDAH	HHN	+01:00	*HAHN AB	HAHN, GE
EDAR	RMS	+01:00	RAMSTEIN AB	LANDSTUHL, GE
EDAS	SEX	+01:00	SEMBACH AB (RUNWAY CLOSED)	SEMBACH, GE
EDBB	THF	+01:00	*TEMPELHOF CENTRAL APT/AIR STATION	BERLN, GE
EDDF	FRA	+01:00	FRANKFURT MAIN IAP	FRANKFURT, GE
EDDH	HAM	+01:00	HAMBURG APT	HAMBURG, GE

EDDN	NUE	+01:00	NURNBERG APT	NURNBERG, GE
EDDS	STR	+01:00	STUTTGART APT	STUTTGART, GE
EDNA	LHN	+01:00	*AHLHORN GAFB	AHLHORN, GE
EDNN	NRV	+01:00	*NORVENICH GAFB	NORVENICH, GE
EDNO	OBG	+01:00	*OLDENBURG GAFB	OLDENBURG, GE
EDSD	LPH	+01:00	*LEIPHEIM GAFB	LEIPHEIM, GE
EDSF	FEL	+01:00	FURSTENFELD-BRUCK AAF	FURSTENFELD-BRUCK, GE
EDSI	IOT	+01:00	*INGOLSTADT AB	INGOLSTADT, GE
EGPK	PIK	+00:00	*PRESTWICK APT	PRESTWICK, SCOTLAND (UK)
EGSS	STN	+00:00	STANSTED APT	STANSTED, UK
EGUL	LKH	+00:00	RAF LAKENHEATH	SUFFOLK, UK
EGUN	MHZ	+00:00	RAF MILDENHALL	SUFFOLK, UK
EGUP	FKH	+00:00	*RAF SCULTHORPE	NORFOLK, UK
EGVG	WOB	+00:00	RAF WOODBRIDGE	SUFFOLK, UK
EGVJ	BWY	+00:00	RAF BENTWATERS	SUFFOLK, UK
EGWZ	AYH	+00:00	RAF ALCONBURY	SUFFOLK, UK
EHSB	SSS	+01:00	*SOESTERBERG RNLAF	UTRECHT, NT
EINN	SNN	+00:00	SHANNON APT	LIMERICK, IR
ENFB	OSL	+01:00	OSLO FORNEBU	OSLO, NO
FAJS	JNB	+02:00	*JAN SMUTS APT	JOHANNESBURG, SF
FHAW	ASI	+00:00	ASCENSION AUX AF	GEORGETOWN ASC, UK
FJDG	NKW	+00:00	DIEGO GARCIA ATOLL	CHAGOS, IO
FTTJ	NDJ	+01:00	N'DJAMENA IAP	N'DJAMENA, CD
FZAA	FIH	+01:00	KINSHASA NDJILI APT	KINSHASA, ZA
GLRB	ROB	+00:00	MONROVIA ROBERTS IAP	MONROVIA, LI
GOOY	DKR	+03:00	DAKAR YOFF	DAKAR, SE
HCMI	BBE	+03:00	BERBERA	BERBERA, SM
HCMM	MGQ	+03:00	PETRELLA APT	MOGADISHU, SM
HECA	CAI	+02:00	CAIRO IAP	CAIRO, EG
HKNA	NBO	+03:00	EMBAKASI APT	NAIROBI, KE
HSSP	PZU	+02:00	PORT SUDAN APT	PORT SUDAN,SU
HSSS	KRT	+02:00	KHARTOUM APT	KHARTOUM SU
KADW	ADW	-05:00	ANDREWS AFB	CAMP SPRINGS, MD
KATL	ATL	-05:00	ATLANTA IAP	ATLANTA, GA
KBDL	BDL	-05:00	BRADLEY IAP	WINDSOR LOCKS, CT
KBGR	BGR	-05:00	BANGOR IAP	BANGOR, ME
KBIX	BIX	-06:00	KEESLER AFB	BILOXI, MS
KBLV	BLV	-06:00	SCOTT AFB	BELLEVILLE, IL
KBNA	BNA	-08:00	NASHVILLE METRO APT	NASHVILLE, TN
KCEF	CEF	-05:00	WESTOVER AFB	CHICOPEE, MA
KCHS	CHS	-05:00	CHARLESTON AFB-IAP	CHARLESTON, SC
KCLT	CLT	-05:00	DOUGLAS IAP	CHARLOTTE, NC
KCOF	COF	-05:00	PATRICK AFB	COCOA BEACH, FL
KCRW	CRW	-05:00	YEAGER APT	CHARLESTON, WV
KCYS	CYS	-07:00	CHEYENNE MUNI APT	CHEYENNE, WY
KDOV	DOV	-05:00	DOVER AFB	DOVER, DE
KDYS	DYS	-05:00	DYESS AFB	ABILENE, TX
KFLL	FLL	-05:00	FT LAUDERDALE INTL	FT LAUDERDALE, FL
KGRK	GRK	-06:00	ROBERT GRAY AAF	KILLEEN, TX
KGSB	GSB	-05:00	SEYMOUR JOHNSON AFB	GOLDSBORO, NC
KILG	ILG	-05:00	NEW CASTLE CO APT	WILMINGTON, DE
KJAN	JAN	-06:00	JACKSON IAP	JACKSON,MS

KJFK	JFK	-05:00	JOHN F KENNEDY IAP	NEW YORK, NY
KLAX	LAX	-08:00	LOS ANGELES IAP	LOS ANGELES, CA
KLFI	LFI	-05:00	LANGLEY AFB	HAMPTON, VA
KLRF	LRF	-06:00	LITTLE ROCK AFB	JACKSONVILLE, AR
KLTS	LIS	-06:00	ALTUS AFB	ALTUS, OK
KMDT	MDT	-05:00	HARRISBURG IAP	MIDDLETOWN, PA
KMFD	MFD	-05:00	MANSFIELD LAHM MUNICIPAL APT	MANSFIELD, OH
KMIA	MIA	-05:00	MIAMI IAP	MIAMI, FL
KMRB	MRB	-05:00	EAST WV REGIONAL APT	MARTINSBURG, WV
KMSP	MSP	-06:00	MPLS-ST PAUL IAP	MINNEAPOLIS, MN
KMTN	MTN	-05:00	MARTIN STATE APT	BALTIMORE, MD
KNBE	NBE	-06:00	DALLAS NAS	DALLAS, TX
KNGU	NGU	-05:00	NORFOLK NAS	NORFOLK, VA
KNIP	NIP	-05:00	JACKSONVILLE NAS	JACKSONVILLE, FL
KNQA	NQA	-06:00	MEMPHIS NAS	MEMPHIS, TN
KOAK	OAK	-08:00	METRO OAKLAND IAP	OAKLAND, CA
KOKC	OKC	-06:00	WILL ROGERS WLD APT	OKLAHOMA CITY, OK
KORF	ORF	-05:00	NORFOLK IAP	NORFOLK, VA
KPHL	PHL	-05:00	PHILADELPHIA IAP	PHILADELPHIA, PA
KPOB	POB	-05:00	POPE AFB	FAYETTEVILLE, NC
KPVD	PVD	-05:00	THEO FRAN GREEN ST	PROVIDENCE, RI
KRIV	RIV	-08:00	MARCH AFB	RIVERSIDE, CA
KSAV	SAV	-05:00	SAVANNAH IAP	SAVANNAH, GA
KSCH	SCH	-05:00	SCHENECTADY CO APT	SCHENECTADY, NY
KSFO	SFO	-08:00	SAN FRANCISCO IAP	SAN FRANCISCO, CA
KSKF	SKF	-06:00	KELLY AFB	SAN ANTONIO, TX
KSTJ	STJ	-06:00	ROSECRANS MEM APT	ST JOSEPH, MO
KSTL	STL	-06:00	LAMBERT-ST LOUIS IAP	ST LOUIS, MO
KSUU	SUU	-08:00	TRAVIS AFB	FAIRFIELD, CA
KSWF	SWF	-05:00	STEWART IAP	NEWBURGH, NY
KTCM	TCM	-08:00	MCCHORD AFB	TACOMA, WA
KTIK	TIK	-06:00	TINKER AFB	OKLAHOMA CITY, OK
KTPA	TPA	-05:00	TAMPA INTL APT	TAMPA, FL
KVNY	VNY	-08:00	VAN NUYS APT	VAN NUYS, CA
KWRI	WRI	-05:00	MCGUIRE AFB	WRIGHTSTOWN, NJ
LCLK	LCA	+02:00	*LARNACA RAFB	LARNACA, CY
LCRA	AKT	+02:00	AKROTIRI RAFB	AKROTIRI, CY
LEMH	MAH	+01:00	MENORCA APT	MAHON, SP
LERT	RTA	+01:00	ROTA NAS	ROTA, SP
LESJ	PMI	+01:00	PALMA DE MALLORCA APT	BALEARIC IS, SP
LETO	TOJ	+01:00	TORREJON DE ARDOZ AB	MADRID, SP
LGAT	ATH	+02:00	*ATHINAI APT	ATHENS, GR
LGIR	VWH	+02:00	IRAKLION AS	IRAKUON (CR), GR
LGSA	SOC	+02:00	SOUDA BAY NAF	KHANIA (CR), GR
LGTS	SKG	+02:00	THESSALONIKI APT	THESSALONIKI, GR
LIBR	BDS	+01:00	BRINDISI/CASALE APT	CAMPO CASALE, IT
LICR	REG	+01:00	*REGGIO CALABRIA ITAB	TITO, IT
LICZ	SIZ	+01:00	SIGONELLA APT	GERBINI (SICILY), IT
LIED	DCU	+01:00	DECIMOMANNU ITAB	DECIMOMANNU, IT
LIEO	OLB	+01:00	OLBIA/COSTA ESMERALDA	OLBIA, IT
LIPA	AVB	+01:00	AVIANO AB	AVIANO, IT
LIPT	VCE	+01:00	VICENZA	VICENZA, IT
LIPX	VRN	+01:00	VILLAFRANCA APT	VERONA, IT
LIPZ	VEN	+01:00	VENEZIA/TESSERA APT	VENICE, IT
LIRN	NAP	+01:00	CAPODICHINO APT	NAPLES,IT
LIRP	PSA	+01:00	SAN GUISTO APT	PISA, IT
LLBG	TLV	+02:00	BEN GURION IAP	TEL AVIV, IS

LPAR	ALA	+00:00	ALVERCA PAFB	ALVERCA, PO
LPLA	LGS	+00:00	LAJES AB	LAJES (AZORES IS), PO
LTAG	ADA	+02:00	INCIRLIK APT	ADANA, TU
LTAT	EHC	+02:00	*ERHAC TUAF	MALATYA, TU
LTBA	YES	+02:00	YESILKOY IAP	ISTANBUL, TU
LTBF	BZI	+02:00	*BALIKESIR APT	BALIKESIR, TU
LTBI	ESK	+02:00	*ESKISEHIR TUAF	ESKISEHIR, TU
LTBL	IGL	+02:00	CIGU TARB	IZMIR, TU
LTCC	DIY	+02:00	*DIYARBAKIR APT	DIYARBAKIR, TU
LTCE	ERZ	+02:00	*EAZURUM APT	ERZURUM, TU
MDSI	SDQ	-04:00	SAN ISIDRO AB	SANTO DOMINGO, DR
MGGT	GUA	-06:00	LA AURORA APT	GUATEMALA CITY, GT
MHLC	LCE	-06:00	GOLOSON IAP	LA CEIBA, HO
MHLM	SAP	-06:00	LA MESA IAP	LA MESA, HO
MHSC	PLA	-06:00	SOTO CANO AB	COMAYAGUA, HO
MHTG	TGU	-06:00	TONCONTIN IAP	TEGUCIGALPA, HO
MKJP	KIN	-05:00	NIIGATA IAP	KINGSTON, JM
MKJT	GDT	-05:00	GRAND TURK AUX FIELD	GRAND TURK IS, UK
MMCZ	CZM	-06:00	COZUMEL INTL	COZUMEL, MX
MMUN	CUN	-06:00	CANCUN INTL	CANCUN, MX
MNMG	MGA	-06:00	LAS MERCEDES APT	MANAGUA, NI
MPHO	HOW	-05:00	HOWARD AB	BALBOA, PN
MPTO	PTY	-05:00	TOCUMEN/TORRIJOS IAP	PANAMA CITY, PN
MROC	OCO	-06:00	EL COCO APT	SAN JOSE, CS
MSSS	SAL	-06:00	ILOPANGO IAP	SAN SALVADOR, ES
MTPP	PAP	-05:00	HAITI IAP	PORT-AU-PRINCE, HA
MUGM	GAO	-05:00	GUANTANAMO BAY NAS	GUANTANAMO, CU
MYGM	GBI	-05:00	GRAND BAHAMA AUX AF	GRAND BAHAMA, BH
MZBZ	BZE	-06:00	BELIZE IAP	BELIZE CITY, BZ
NSTU	PPG	-11:00	PAGO PAGO IAP	PAGO PAGO, AS
NZCH	CHC	+12:00	CHRISTCHURCH IAP	CHRISTCHURCH, NZ
OBBI	BAH	+03:00	BAHRAIN IAP	AL MUHARRAQ, BA
OEDR	DHA	+03:00	DHAHRAN IAP	DHAHRAN, SA
OEJB		+03:00	JUBAIL	JUBAIL, SA
OEJD	JED	+03:00	JIDDAH IAP	JIDDAH, SA
OEJF	KFJ	+03:00	KING FAISAL NB	JIDDAH, SA
OEJN	JDW	+03:00	KING ABDUL AZIZ IAP	JIDDAH, SA
OEKK		+03:00	KING KAHLID MILIT	KING KHALID MILIT, SA
OEKM	KAI	+03:00	KHAMS MUSHAIT AB	ABHA, SA
OERY	RUH	+03:00	RIYADH IAP	RIYADH, SA
OETB	TUU	+03:00	KING FAISAL AB	TABUK, SA
OETF	TIF	+03:00	TAIF APT	TAIF, SA
OJAF	AMM	+02:00	KING ABDULLAH AB	AMMAN, JR
OOMA	MSH	+04:00	MASIRAH OAFB	MASIRAH, OM
OOMS	SBE	+04:00	SEEB IAP	MUSCAT, OM
OOTH	TTH	+04:00	THUMRAIT OAFB	MIDWAY, OM
PACD	CDB	-09:00	COLD BAY APT	COLD BAY, AK
PACZ	CZF	-09:00	CAPE ROMANZOF AFS	CAPE ROMANZOF, AK
PADK	ADK	-10:00	ADAK ISLAND NS	ADAK, AK
PAED	EDF	-09:00	ELMENDORF AFB	ANCHORAGE, AK
PAEH	EHM	-09:00	CAPE NEWENHAM AFS	CAPE NEWENHAM, AK
PAEI	EIL	-09:00	EIELSON AFB	FAIRBANKS, AK
PAFA	FAI	-09:00	FAIRBANKS IAP	FAIRBANKS, AK
PAFY	FYU	-09:00	FORT YUKON APT	FORT YUKON, AK
PAGA	GAL	-09:00	GALENA APT	GALENA, AK
PAHT	AHT	-09:00	AMCHITKA ISL APT	ANCHITKA, AK
PAIM	UTO	-09:00	INDIAN MOUNTAIN AFS	UTOPIA CREEK, AK
PAKN	AKN	-09:00	KING SALMON APT	KING SALMON, AK

PALU	LUR	-09:00	CAPE LISBURNE AFS	CAPE LISBURNE, AK
PANC	ANO	-09:00	ANCHORAGE IAP	ANCHORAGE, AK
PAOT	OTZ	-09:00	RALPH WIEN MEMORIAL	KOTZEBUE, AK
PASV	SVW	-10:00	SPARREVOHN AFS	SPARREVOHN, AK
PASY	SYA	-10:00.	EARECKSON AFB	EARECKSON, AK
PATC	TNC	-09:00.	TIN CITY AFS	TIN CITY, AK
PATL	TLJ	-09:00	TATALINA AFS	TATALINA, AK
PGSN	SPN	+10:00	SAIPAN IAP	MARIANA IS(SAIPAN), US
PGUA	UAM	+10:00	ANDERSEN AFB	GUAM MARIANAS, GU
PGUM	GUM	+10:00	AGANA NAS	BREWER FIELD, GU
PHIK	HIK	-10:00	HICKAM AFB	HONOLULU, HI
PHNL	HNL	-10:00	HONOLULU IAP	HONOLULU, HI
PJON	JON	-10:00	JOHNSTON ATOLL AFB	JOHNSTON IS, JO
PKMA	ENT	+12:00	ENEWETAK AUX AF	MARSHALL IS, MI
PKWA	KWA	-12:00	BUCHOLZ AAF KMR	KWAJALEIN ATOLL, MI
PMDY	MDY	-11:00	MIDWAY NS	MIDWAY IS, MW
PTKK	TKK	+10:00	TRUK IAP	MOEN IS, FM
PTPN	PNI	+11:00	PONAPE IAP	PONAPE, FM
PTRO	ROR	+09:00	BABELTHUAP	PALAU IS, TTPI
PTTK	KSA	+09:00	KOSRAE APT	KOSRAE, FM
PTYA	YAP	+10:00	YAP IAP	YAP CAROLINE IS, FM
PWAK	AWK	+12:00	WAKE ISLAND AFB	WAKE IS, WK
RJAM	MUS	+10:00	MINAMI TORISHIMA APT	MARCUS IS, JA
RJAW	IWO	+09:00	IWO JIMA AB	IWO JIMA IS, JA
JCB	OBO	+09:00	OBIHIRO APT	OBIHIRO HOKKAIDO, JA
JCJ	CTS	+09:00	CHITOSE APT	SAPPORO, JA
JCK	KUH	+09:00	KUSHIRO AB	KUSHIRO HOKKAIDO, JA
JFF	FUK	+09:00	FUKUOKA/ITAZUKE	KYUSHU IS, JA
JOI	IWA	+09:00	IWAKUNI MCAS	HONSHU IS, JA
JSM	MSJ	+09:00	MISAWA AB	HONSHU IS, JA
JTA	NJA	+09:00	ATSUGI NAS	HONSHU IS, JA
JTY	OKO	+09:00	YOKOTA AB	FUSSA, JA
KJJ	KWJ	+09:00	*KWANG JU ROKAFB	KWANG JU, RK
KJK	KUZ	+09:00	KUNSAN AB	KUNSAN, RK
KPC	CJU	+09:00	CHEJU IAP	CHEJU DO IS, RK
KPK	KHE	+09:00	KIMHAE IAP	PUSAN, RK
KSO	OSN	+09:00	OSAN AB	OSAN, RK
KSS	SEL	+09:00	KIMPO IAP	SEOUl, RK
KSW	HLV	+09:00	*SUWON ROKAFB	SUWON, RK
KTN	TAE	+09:00	*TAEGU AB	TAEGU, RK
OAH	OKA	+09:00	NAHA APT	OKINAWA IS, JA
ODN	DNA	+09:00	KADENA AB	OKINAWA IS, JA
ADP	PAL	-03:00	EL PALOMAR AF	BUENOS AIRES, AG
AEZ	BUE	-03:00	EZEIZA APT	BUENOS AIRES, AG
BBR	BSB	-03:00	BRASILIA APT	BRASILIA, BR
BGL	RIO	-03:00	RIO DE JANEIRO IAP	RIO DE JANEIRO, BR
CEL	SCL	-04:00	PUDAHEL APT	SANTIAGO, CH
EQU	UIO	-05:00	MARISCAL SUCRE APT	QUITO, EC
GAS	ASU	-04:00	PRES STROESSNER APT	ASUNCION, PG
KBO	BOG	-05:00	EL DORADO IAP	BOGOTA, CL
KCG	CTG	-05:00	CARTAGENA APT	CARTAGENA, CL
LP	LPB	-04:00	EL ALTO APT	LA PAZ, BO
MZY	PBM	-03:00	ZANDERY APT	PARAMARIBO, SR
PIM	LIM	-05:00	JORGE CHAVEZ IAP	LIMA, PE
UMU	MVD	-03:00	CARRASCO IAP	MONTEVIDEO, UG
VBL	ELR	-04:00	EL LIBERATADOR AB	NEGRO, VE
VMI	MIQ	-04:00	MAIQUETA APT	MAIQUETA, VE
YTM	GEO	-03:00	TIMEHRI IAP	GEORGETOWN, GY

TAPA	SJH	-04:00	V C BIRD IAP	ST JOHNS, AN
TBPB	BGI	-04:00	GRANTLEY ADAMS IAP	BRIDGETOWN, BB
TISX	STX	-04;00	ALEX HAMILTON APT	ST CROIX, VI
TJNR	NRR	-04:00	ROOSEVELT ROADS NAS	ROOSEVELT ROADS, PR
TXKF	BOA	-04:00	BERMUDA NAS	HAMILTON, BM
VTBD	BKK	+07:00	DON MUANG APT	BANGKOK, TH
VTBU	VBU	+07:00	U-TAPAO RTN	BAN U-TAPAO, TH
WIIH	DJK	+07:00	JAKARTA APT	JAKART, IE
WMKK	MKK	+08:00.	KUALA LUMPUR IAP	SUBANG, MA
WSAP	SGP	+08:00	PAYA LEBARRSAF	SINGAPORE, SG

DECODE (LI vs ICAO)

LI	ICAO	LST	NAME	LOCATION
ADA	LTAG	+02:00	INCIRLIK APT	ADANA, TU
ADK	PADK	-10:00	ADAK ISLAND NS	ADAK, AK
ADL	AAAD	+09:30	ADELAIDE APT	WEST TORRENS, AU
ADW	KADW	-05:00	ANDREWS AFB	CAMP SPRINGS, MD
AHT	PAHT	-09:00	AMCHITKA	AMCHITKA, AK
AKN	PAKN	-09:00	KING SALMON APT	KING SALMON, AK
AKT	LCRA	+02:00	AKROTIRI RAFB	AKROTIRI, CY
ALA	LPAR	+00:00	ALVERCA PAFB	ALVERCA, PO
AMM	OJAF	+02:00	KING ABDULLAH AB	AMMAN, JR
ANC	PANC	-09:00	ANCHORAGE IAP	ANCHORAGE, AK
ASI	FHAW	+00:00	ASCENSION AUX AF	GEORGETOWN ASC, UK
ASP	ABAS	+09:30	ALICE SPRINGS RAAFB	ALICE SPRINGS, AU
ASU	SGAS	-04:00	PRES STROESSNER APT	ASUNCION, PG
ATH	LGAT	+02:00	*ATHINA IAP	ATHENS, GR
ATL	KATL	-05:00	ATLANTA IAP	ATLANTA, GA
AVB	LIPA	+01:00	AVIANO AB	AVIANO, IT
AWK	PWAK	+12:00	WAKE ISLAND AFB	WAKE IS, WK
AYH	EGWZ	+00:00	RAF ALCONBURY	SUFFOLK, UK
BAH	OBBI	+03:00	BAHRAIN IAP	AL MUHARRAQ, BA
BBE	HCMI	+03:00.	BERBERA	BERBERA, SM
BDA	TXKF	-04:00	BERMUDA NAS	HAMLTON, BM
BDL	KBDL	-05:00	BRADLEY IAP	WINDSOR LOCKS, CT
BDS	LIBR	+01:00	BRINDISI/CASALE APT	CAMPO CASALE, IT
BGI	TBPB	-04:00	GRANTLEY ADAMS IAP	BRIDGETOWN, BB
BGR	KBGR	-05:00	BANGOR IAP	BANGOR, ME
BIX	KBIX	-06:00	KEESLER AFB	BILOXI, MS
BKK	VTBD	+07:00	DON MUANG APT	BANGKOK, TH
BLV	KBLV	-06:00	SCOTT AFB	BELLEVILLE, IL
BNA	KBNA	-06:00	NASHVILLE METRO APT	NASHVILLE, TN
BOG	SKBO	-05:00	EL DORADO IAP	BOGOTA, CL
BRU	EBBR	+01:00	BRUSSELS NATL	BRUSSELS, BE
BSB	SBBR	-03:00	BRASILIA APT	BRASILIA, BR
BUE	SAEZ	-03:00	EZEIZA APT	BUENOS AIRES, AG
BWY	EGVJ	+00:00	RAF BENTWATERS	SUFFOLK, UK
BZE	MZBZ	-06:00	BELIZE IAP	BELIZE CITY, BZ
BZI	LTBF	+02:00	*BALIKESIR APT	BALIKESIR, TU
CAI	HECA	+02:00	CAIRO IAP	CAIRO, EG
CDB	PACD	-09:00	COLD BAY APT	COLD BAY, AK
CEF	KCEF	-05:00	WESTOVER AFB	CHICOPEE, MA
CGY	RPWL	+09:00	CAGAYAN DE ORO APT	MINDANAO IS, RP
CHC	NZCH	+12:00	CHRISTCHURCH IAP	CHRISTCHURCH, NZ
CHS	KCHS	-05:00	CHARLESTON AFB	CHARLESTON, SC

CJU	RKPC	+09:00	CHEJU IAP	CHEJU DO IS, RK
CLT	KCLT	-05:00	DOUGLAS IAP	CHARLOTTE, NC
COF	KCOF	-05:00	PATRICK AFB	COCOA BEACH, FL
CRW	KCRW	-05:00	YEAGER APT	CHARLESTON, WV
CTG	SKCG	-05:00	CARTAGENA APT	CARTAGENA, CL
CTS	RJCJ	+09:00	CHITOSE APT	SAPPORO, JA
CUN	MMUN	-06:00	CANCUN INTL	CANCUN, MX
CYS	KCYS	-07:00	CHEYENNE MUNI APT	CHEYENNE, WY
CZF	PACZ	-09:00	CAPE ROMANZOF AFS	CAPE ROMANZOF, AK
CZM	MMCZ	-06:00	COZUMEL INTL	COZUMEL, MX
DCU	LIED	+01:00	*DECIMOMANNU ITAB	DECIMOMANNU, IT
DHA	OEDR	+03:00	DHAHRAN IAP	DHAHRAN, SA
DIV	LTCC	+02:00	DIYARBAKIR APT	DIYARBAKIR, TU
DJK	WIIH	+07:00	JAKARTA APT	JAKARTA, IE
DKR	GOOY	+03:00	DAKAR YOFF	DAKAR, SE
DNA	RODN	+09:00	KADENA AFB	OKINAWA, JA
DOV	KDOV	-05:00	DOVER AFB	DOVER, DE
DYS	KDYS	-06:00	DYESS AFB	ABILENE, TX
EDF	PAED	-09:00	ELMENDORF AFB	ANCHORAGE, AK
EHC	LTAT	+02:00	*EHRAC TUAF	MALATYA, TU
EHM	PAEH	-09:00	CAPE NEWENHAM AFS	CAPE NEWENHAM, AK
EIL	PAEI	-09:00	EIELSON AFB	FAIRBANKS, AK
ELR	SVBL	-04:00	EL LIBERATADOR AB	NEGRO, VE
ENT	PKMA	+12:00	ENEWETAK AUX AF	MARSHALL IS, MI
ERZ	LTCE	+02:00	*ERZURUM APT	ERZURUM, TU
ESB	LTAC	+02:00	*ESENBOGA APT	ANKARA, TU
ESK	LTBI	+02:00	*ESKISEHIR TUAF	ESKISEHIR, TU
FM	PAFA	-00:00	FAIRBANKS IAP	FAIRBANKS, AK
FEL	EDSF	+01:00	FURSTENFELD- BRUCK AAF	FURSTENFELD- BRUCK, GE
FIH	FZAA	+01:00	KINSHASA NDJILI APT	KINSHASA, ZA
FKH	EGUP	+00:00	*RAF SCULTHORPE	NORFOLK, UK
FLL	KFLL	-05:00	FT LAUDERDALE INTL	FT LAUDERDALE, FL
FRA	EDOF	+01:00	FRANKFURT MAIN IAP	FRANKFURT, GE
FRF	EDAF	+01:00	RHEIN-MAIN AB	FRANKFURT, GE
FUK	RJFF	+09:00	FUKUOKA/ITAZUKE	KYUSHU IS, JA
FYU	PAFY	-09:00	FORT YUKON APT	FORT YUKON, AK
GAL	PAGA	-09:00	GALENA APT	GALENA, AK
GAO	MUGM	-05:00	GUANTANAMO BAY NAS	GUANTANAMO, CU
GBI	MYGM	-05:00	GRAND BAHAMA AUX AF	GRAND BAHAMA, BH
GDT	MKJT	-05:00	GRAND TURK AUX FIELD	GRAND TURK IS, UK
GEO	SYTM	-03:00	TIMEHRI IAP	GEORGETOWN, GY
GRK	KGRK	-06:00	ROBERT GRAY AAF	KILLEEN, TX
GSB	KGSB	-05:00	SEYMOUR JOHNSON AFB	GOLDSBORO, NC
GUA	MGGT	-06:00	LA AURORA APT	GUATEMALA CITY, GT
GUM	PGUM	+10:00	AGANA NAS	BREWER FIELD, GU
HAM	EDOH	+01:00	HAMBURG APT	HAMBURG, GE
HHN	EDAH	+01:00	*HAHN AB	HAHN, GE
HIK	PHIK	-10:00	HICKAM AFB	HONOLULU, HI
HIR	AGGH	+11:00	HENDERSON IAP	HONIARA, SI
HLV	RKSW	+09:00	SUWON ROKAFB	SUWON, RK
HNL	PHNL	-10:00	HONOLULU IAP	HONOLULU, HI
HOW	MPHO	-05:00	HOWARD AB	BALBOA, PN
HRT		-06:00	HURLBURT FIELD	MARY ESTHER, FL
IDT	EDSI	+01:00	*INGOLSTADT AB	INGOLSTADT, GE
IGL	LTBL	+02:00	CIGLI TUAF	IZMIR, TU
ILG	KILG	-05:00	NEW CASTLE CO APT	WILMINGTON, DE
IWA	RJOI	+09:00	IWAKUNI MCAS	HONSHU IS, JA

IWO	RJAW	+09:00	IWO JIMA AB	IWO JIMA IS, JA
JAN	KJAN	-06:00	JACKSON UAP	JACKSON, MS
JDW	OEJN	+03:00	KING ABDUL AZIZ IAP	JIDDAH, SA
JED	OEJN	+03:00	JIDDAH IAP	JIDDAH, SA
JFK	KJFK	-05:00	JOHN F. KENNEDY IAP	NEW YORK, NY
JNB	FAJS	+02:00	*JAN SMUTS APT	JOHANNESBURG, SF
JON	PJON	-10:00	JOHNSTON ATOLL AFB	JOHNSTON IS, JO
KAI	OEKM	+03:00	KHAMIS MUSHAIT AB	ABHA, SA
KEF	BIKF	+00:00	KEFLAVIK APT	KEFLAVIK, IC
KFJ	OEJF	+03:00	KING FAISAL NB	JIDDAH, SA
KHE	RKPK	+09:00	*KIMHAE IAP	PUSAN, RK
KIN	MKJP	-05:00	NIIGATA IAP	KINGSTON, JM
KRT	HSSS	+02:00	KHARTOUM APT	KHARTOUM, SU
KSA	PTTK	+09:00	KOSRAE APT	KOSRAE, FM
KUH	RJCK	+09:00	KUSHIRO AB	KUSHIRO HOKKAIDO, JA
KUZ	RKJK	+09:00	KUNSAN AB	KUNSAN, RK
KWA	PKWA	-12:00	BUCHOLZ AAF KMR	KWAJALEIN ATOLL, MI
LAX	KLAX	-08:00	LOS ANGELES IAP	LOS ANGELES, CA
LCE	MHLC	-06:00	GOLDSON IAP	LA CEIBA, HO
LEA	APLM	+08:00	*LEARMONTH RAAFB	EXMOUTH GULF, AU
LFI	KLFI	-05:00	LANGLEY AFA	HAMPTON, VA
LGS	LPLA	+00:00	LAJES AB	LAJES (AZORES IS), PO
LHN	EDNA	+01:00	*AHLHORN GAFB	ANLHORN, GE
LIM	SPIM	-05:00	JORGE CHAVEZ IAP	LIMA, PE
LKH	EGUL	+00:00	RAF LAKENHEATH	SUFFOLK, UK
LPB	SLLP	-04:00	EL ALTO APT	LA PAZ, BO
LPH	EDSD	+01:00	*LEIPHEIM GAFB	LEIPHEIM, GE
LRF	KLRF	-06:00	LITTLE ROCK AFB	JACKSONVILLE, AR
LTS	KLTS	-06:00	ALTUS AFB	ALTUS. OK
LUR	PALU	-09:00	CAPE LISBURNE AFS	CAPE LISBURNE, AK
MAH	LEMH	+01:00	MENORCA APT	MAHON, SP
MDT	KMDT	-05:00	HARRISBURG LAP	MIDDLETOWN, PA
MDY	PMDY	-11:00	MIDWAY NS	MIDWAY IS,MW
MFD	KMFD	-05:00	MANSFIELD LAHM MUNICIPAL APT	MANSFIELD, OH
MGA	MNMG	-06:00	LAS MERCEDES APT	MANAGUA, NI
MGQ	HCMM	+03:00	PETRELLA APT	MOGADISHU, SM
MHZ	EGUN	+00:00	RAF MILDENHALL	SUFFOLK, UK
MIA	KMIA	-05:00	MIAMI IAP	MIAMI, FL
MIQ	SVMI	-04:00	MAIQUETA APT	MAIQUETA, VE
MKK	WMKK	+08:00	KUALA LUMPUR IAP	SUBANG, MA
MRB	KMRB	-05:00	EAST WV REGIONAL APT	MARTINSBURG, WV
MSH	OOMA	+04:00	MASIRAH OAFB	MASIRAH, OM
MSJ	RJSM	+09:00	MISAWA AB	HONSHU IS, JA
MSP	KMSP	-06:00	MPLS-ST PAUL IAP	MINNEAPOUS, MN
MTN	KMTN	-05:00	MARTIN STATE APT	BALTIMORE, MD
MUS	RJAM	+10:00	MINAMI TORISHIMA APT	MARCUS IS, JA
MVD	SUMU	-03:00	CARRASCO IAP	MONTEVIDEO, UG
NAP	LIRN	+01:00	CAPODICHINO APT	NAPLES, IT
NBE	KNBE	-06:00	DALLAS NAS	DALLAS, TX
NBO	HKNA	+03:00	EMBAKASI APT	NAIROBI, KE
NDJ	FTTJ	+01:00	N'DJAMENA IAP	N'DJAMENA, CD
NGU	KNCU	-05:00	NORFOLK NAS	NORFOLK, VA
NIM	DARN	+01:00	NIAMEY IAP	NIAMEY, NG
NIP	KNIP	-05:00	JACKSONVILLE NAS	JACKSONVILLE, FL
NJA	RJTA	+09:00	ATSUGI NAS	HONSHU IS, JA
NKW	FJDG	+00:00	DIEGO GARCIA ATOLL	CHAGOS IS, IO
NOA	KNOA	-06:00	MEMPHIS NAS	MEMPHIS, TN

NRR	TJNR	-04:00	ROOSEVELT ROADS NAS	ROOSEVELT ROADS, PR
NRV	EDNN	+01:00	*NORVENICH GAFB	NORVENICH, GE
NUE	EDDN	+01:00	NURNBERG APT	NURNBERG, GE
OAK	KOAK	-08:00	METRO OAKLAND IAP	OAKLAND, CA
OBG	EDNO	+01:00	*OLDENBURG GAFB	OLDENBURG, GE
OBO	RJCB	+09:00	OBHIRO APT	OBHIRO HOKKAIDO, JA
OCO	MROC	-06:00	EL COCO APT	SAN JOSE,CS
OKA	ROAH	+09:00	NAHA APT	OKINAWA IS, JA
OKC	KOKC	-06:00	WILL ROGERS WLD APT	OKLAHOMA CITY, OK
OKO	RJTY	+09:00	YOKOTA AB	FUSSA, JA
OLB	LIEO	+01:00	OLBIA/COSTA ESMERALDA	OLIBIA, IT
ORF	KORF	-05:00	NORFOLK IAP	NORFOLK, VA
OSL	ENFB	+01:00	OSLO FORNEBU	OSLO, NO
OSN	RKSO	+09:00	OSAN AB	OSAN, RK
OTZ	PAOT	-09:00	RALPH WEIN MEMORIAL	KOTZEBUE, AK
PAL	SADP	-03:00	EL PALOMAR AF	BUENOS AIRES, AG
PAP	MTPP	-05:00	HAITI IAP	PORT-AU-PRINCE, HA
PBM	SMZY	-03:00	ZANDEREY APT	PARAMARIBO, SR
PHL	KPHL	-05:00	PHILADELPHIA IAP	PHILADELPHIA, PA
PIK	EGPK	+00:00	*PRESTWICK APT	PRESTWICK (SCOTL), UK
PLA	MHSC	-06:00	SOTO CANO AB	COMAYAGUA, HO
PMI	LESJ	+01:00	PALMA DE MALLORCA APT	BALEARIC IS, SP
PNI	PTPN	+11:00	PONAPE IAP	PONAPE, FM
POB	KPOB	-05:00	POPE AFB	FAYETTEVILLE, NC
PPG	NSTU	-11:00	PAGO PAGO IAP	PAGO PAGO, AS
PSA	LIRP	+01:00	SAN GUISTO APT	PISA, IT
PTY	MPTO	-05:00	TOCUMEN/TORRIJOS IAP	PANAMA CITY, PN
PVD	KPVD	-05:00	THEO FRAN GREEN ST	PROVIDENCE, RI
PZU	HSSP	+02:00	PORT SUDAN APT	PORT SUDAN, SU
RCM	ASRI	+10:00	RICHMOND RAAFB	RICHMOND, AU
REG	LICR	+01:00	*REGGIO CALABRIA ITAB	TITO, IT
RIO	SBGL	-03:00	RIO DE JANEIRO IAP	RIO DE JANEIRO, BR
RIV	KRIV	-08:00	MARCH AFB	RIVERSIDE, CA
RMS	EDAR	+01:00	RAMSTEIN AB	LANDSTUHL, GE
ROB	GLRB	+00:00	MONROVIA ROBERTS IAP	MONROVIA, LI
ROR	PTRO	+09:00	BABELTHUAP	PALAU IS, TTPI
RTA	LERT	+01:00	ROTA NAS	ROTA, SP
RUH	OERY	+03:00	RIYADH IAP	RIYADH, SA
SAL	MSSS	-06:00	ILOPANGO IAP	SAN SALVADOR, ES
SAP	MHLM	-06:00	LA MESA IAP	LA MESA, HO
SAV	KSAV	-05:00	SAVANNAH IAP	SAVANNAH, GA
SBE	OOMS	+04:00	SEEB IAP	MUSCAT, OM
SCH	KSCH	-05:00	SCHENECTADY CO APT	SCHENECTADY, NY
SCL	SCEL	-04:00	PUDAHEL APT	SANTIAGO, CH
SDQ	MDSI	-04:00	SAN ISIDRO AB	SANTO DOMINGO, DR
SEL	RKSS	+09:00	KIMPO IAP	SEOUL, RK
SEX	EDAS	+01:00	SEMBACH AB (RUNWAY CLOSED)	SEMBACH, GE
SFJ	BGSF	-03:00	*SONDRESTROM AB	SONDRESTROM GL(DN)
SF0	KSFO	-08:00	SAN FRANCISCO IAP	SAN FRANCISCO, CA
SGP	WSAP	+09:00	RSAF PAYA LEBAR	SINGAPORE, SG
SIZ	LICZ	+01:00	SIGONELLA APT	GERBINI (SICILY), IT
SJH	TAPA	-04:00	V C BIRD IAP	ST JOHNS, AN
SKF	KSKF	-06:00	KELLY AFB	SAN ANTONIO, TX
SKG	LGTS	+02:00	THESSALONIKI APT	THESSALONIKI, GR
SNN	EINN	+00:00	SHANNON APT	LIMERICK, IR
SOC	LGSA	+02:00	SOUDA BAY NAF	KHANIA (CR), GR
SPN	PGSN	+10:00	SAIPAN IAP	MARIANA IS(SAIPAN), US

SSS	EHSB	+01:00	*SOESTERBERG RNLAF	UTRECHT, NT
STJ	KSTJ	-06:00	ROSECRANS MEM APT	ST JOSEPH, MO
STL	KSTL	-06:00	LAMBERT-ST LOUIS IAP	ST LOUIS, MO
STN	EGSS	+03:00	STANSTED APT	STANSTED, UK
STR	EDDS	+01:00	STUTTGART APT	STUTTGART, GE
STX	TISX	-04:00	ALEX HAMILTON APT	ST CROIX, VI
SUU	KSUU	-08:00	TRAVIS AFB	FAIRFIELD, CA
SVW	PASV	-09:00	SPARREVOHN AFS	SPARREVOHN, AK
SWF	KSWF	-05:00	STEWART IAP	NEWBURGH, NY
SYA	PASY	-10:00	EARECKSON AFB	SHEMYA, AK
SYD	ASSY	+10:00	SYDNEY IAP	SYDNEY, AU
TAE	RKTN	+09:00	TAEGU AB	TAEGU, RK
TCM	KTCM	-08:00	MCCHORD AFB	TACOMA, WA
TGU	MHTG	-06:00	TONCONTIN IAP	TEGUCIGALPA, HO
THF	EDBB	+01:00	*TEMPLEHOF CENTRAL APT/AIR STATION	BERLIN, GE
THU	BGTL	-04:00	THULE AB	THULE, GL(DN)
TIF	OETF	+03:00	TAIF APT	TAIF, SA
TIK	KTIK	-06:00	TINKER AFB	OKLAHOMA CITY, OK
TKK	PTKK	+09:00	TRUK APT	MOEN IS, FM
TLJ	PATL	-09:00	TATALINA AFS	TATALINA, AK
TLV	LLBG	+02:00	BEN GURION IAP	TEL AVIV, IS
TNC	PATC	-09:00	TIN CITY AFS	TIN CITY, AK
TOJ	LETO	+01:00	TORREJON DE ARDOZ AB	MADRID, SP
TPA	KTPA	-05:00	TAMPA INTL APT	TAMPA, FL
TTH	OOTH	+04:00	THUMRAIT OAFB	MIDWAY, OM
TUU	OETB	+03:00	KING FAISAL AB	TABUK, SA
UAM	PGUA	+10:00	ANDERSEN ARB	GUAM MARIANAS, GU
UIO	SEQU	-05:00	MARISCAL SUCRE APT	QUITO, EC
UMR	APWR	+09:30	WOOMERA AIR STATION	WOOMERA, AU
UTO	PAIM	-09:00	INDIAN MOUNTAIN AFS	UTOPIA CREEK, AK
VBU	VTBU	+07:00	U-TOPAO RTN	BAN U-TOPAO, TH
VCE	LIPT	+01:00	VICENZA	VICENZA, IT
VEN	LIPZ	+01:00	VENEZIA/TESSERA APT	VENICE, IT
VNY	KYNY	-08:00	VAN NUYS APT	VAN NUYS, CA
VRN	LIPX	+01:00	VILLAFRANCA APT	VERONA, IT
VWH	LGIR	+02:00	*IRAKLION AS	IRAKLION (CR), GR
WOB	EGVG	+00:00	RAF WOODBRIDGE	SUFFOLK, UK
WRI	KWRI	-05:00	MCGUIRE AFB	WRIGHTSTOWN, NJ
YAP	PTYA	+10:00	YAP IAP	YAP CAROLINE IS, FM
YAW	CYAW	-04:00	HALIFAX SHEARWATER	HALIFAX (NS), CN
YES	LTBA	+02:00	YESILKOY IAP	ISTANBUL, TU
YQX	CYQX	-03:30	GANDER INTN	NEWFOUNDLAND, ON
YYR	CYYR	-04:00	*GOOSE BAY AB	GOOSE BAY NFLD, CN
YYT	CYYT	-03:30	ST JOHNS APT	ST JOHNS, CN

*** US PORTION CLOSED**

APPENDIX C: JULIAN DATE CALENDARS AND MILITARY (24 HOUR) CLOCK

Julian Date Calendar: The following tables will be used to convert the Gregorian (official United States Commerce Calendar) calendar dates to a three numerical digit Julian Date. The first Julian Date Calendar table is a perpetual calendar for all years except leap years. The second Julian Date Calendar table is a perpetual calendar for leap years only, such as 1996, 2000, 2004, etc.

How to use the Julian Date Calendars: The calendars are constructed in a matrix with the days of the month in the left hand column from 1-31. The months are displayed in rows from left to right starting with January and continuing through each month to December on the extreme right.

If you made an application for Space-A travel on 15 April 1994 (a non-leap year), then read down the date column to 15 and across to the right to April where you find the number 105, or the 105th day in the Julian Date Calendar. If this were a leap year, i.e., 15 April 1996, check the number-it would be the 106th day. Please try to convert some sample dates (birthdays, holidays, etc.) until you are comfortable using this system.

The entire Julian Date is constructed using the last two digits of the calendar year, i.e., 1994 is 94, 105 for 15 April. The last four digits of the Julian Date are taken from the **twenty-four hour clock,** frequently **known as the military clock,** or time where the hour is a period of time equal to one twenty-fourth of a mean solar or civil day and equivalent to sixty minutes. The Julian Day is divided into a series of twenty-four hours from midnight to midnight. See the table below.

Conventional Clock	Military Clock
12am (midnight)	2400 hours
1am	0100 hours
2am	0200 hours
3am	0300 hours
4am	0400 hours
5am	0500 hours
6am	0600 hours
7am	0700 hours
8am	0800 hours
9am	0900 hours

10am	1000 hours
11am	1100 hours
12pm (**noon**)	1200 hours
1pm	1300 hours
2pm	1400 hours
3pm	1500 hours
4pm	1600 hours
5pm	1700 hours
6pm	1800 hours
7pm	1900 hours
8pm	2000 hours
9pm	2100 hours
10pm	2200 hours
11pm	2300 hours

So, if you apply for Space-A air travel at 2:45pm on April 15, 1994, your complete Julian Date and time would be 94 105 1445, or the year 1994, 105th Day, 14th hour and 45 minutes.

- NOTES -

APPENDIX C, continued: JULIAN DATE CALENDARS AND MILITARY (24-HOUR) CLOCK

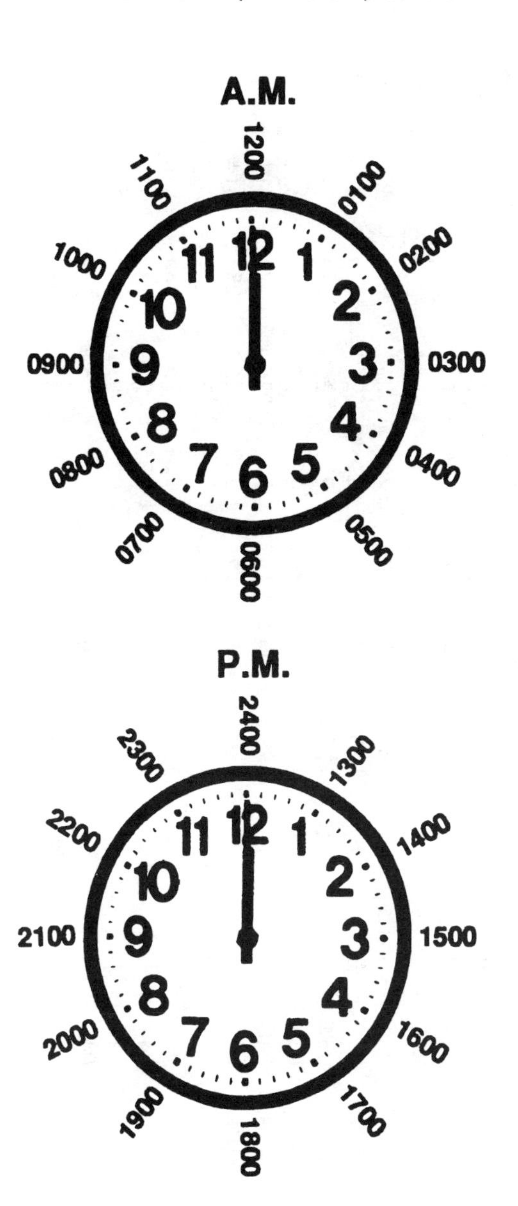

APPENDIX C, continued: JULIAN DATE CALENDAR (NON-LEAP YEAR)

DAY	JAN	FEB	MAR	APR	MAY	JUN	JUL	AUG	SEP	OCT	NOV	DEC
1	001	032	060	091	121	152	182	213	244	274	305	335
2	002	033	061	092	122	153	183	214	245	275	306	336
3	003	034	062	093	123	154	184	215	246	276	307	337
4	004	035	063	094	124	155	185	216	247	277	308	338
5	005	036	064	095	125	156	186	217	248	278	309	339
6	006	037	065	096	126	157	187	218	249	279	310	340
7	007	038	066	097	127	158	188	219	250	280	311	341
8	008	039	067	098	128	159	189	220	251	281	312	342
9	009	040	068	099	129	160	190	221	252	282	313	343
10	010	041	069	100	130	161	191	222	253	283	314	344
11	011	042	070	101	131	162	192	223	254	284	315	345
12	012	043	071	102	132	163	193	224	255	285	316	346
13	013	044	072	103	133	164	194	225	256	286	317	347
14	014	045	073	104	134	165	195	226	257	287	318	348
15	015	046	074	105	135	166	196	227	258	288	319	349
16	016	047	075	106	136	167	197	228	259	289	320	350

APPENDIX C, continued

DAY	JAN	FEB	MAR	APR	MAY	JUN	JUL	AUG	SEP	OCT	NOV	DEC
17	017	048	076	107	137	168	198	229	260	290	321	351
18	018	049	077	108	138	169	199	230	261	291	322	352
19	019	050	078	109	139	170	200	231	262	292	323	353
20	020	051	079	110	140	171	201	232	263	293	324	354
21	021	052	080	111	141	172	202	233	264	294	325	355
22	022	053	081	112	142	173	203	234	265	295	326	356
23	023	054	082	113	143	174	204	235	266	296	327	357
24	024	055	083	114	144	175	205	236	267	297	328	358
25	025	056	084	115	145	176	206	237	268	298	329	359
26	026	057	085	116	146	177	207	238	269	299	330	360
27	027	058	086	117	147	178	208	239	270	300	331	361
28	028	059	087	118	148	179	209	240	271	301	332	362
29	029		088	119	149	180	210	241	272	302	333	363
30	030		089	120	150	181	211	242	273	303	334	364
31	031		090		151		212	243		304		365

APPENDIX C, continued: JULIAN DATE CALENDAR (LEAP YEAR)

DAY	JAN	FEB	MAR	APR	MAY	JUN	JUL	AUG	SEP	OCT	NOV	DEC
01	001	032	061	092	122	153	183	214	245	275	306	336
02	002	033	062	093	123	154	184	215	246	276	307	337
03	003	034	063	094	124	155	185	216	247	277	308	338
04	004	035	064	095	125	156	186	217	248	278	309	339
05	005	036	065	096	12.6	157	187	218	249	279	310	340
06	006	037	066	097	127	158	188	219	250	280	311	341
07	007	038	067	098	128	159	189	220	251	281	312	342
08	008	039	068	099	129	160	190	221	252	282	313	343
09	009	040	069	100	130	161	191	222	253	283	314	344
10	010	041	070	101	131	162	192	223	254	284	315	345
11	011	042	071	102	132	163	193	224	255	285	316	346
12	012	043	072	103	133	164	194	225	256	286	317	347
13	013	044	073	104	134	165	195	226	257	287	318	348
14	014	045	074	105	135	166	196	227	258	288	319	349
15	015	046	075	106	136	167	197	228	259	289	320	350
16	016	047	076	107	137	168	198	229	260	290	321	351

APPENDIX C, continued

DAY	JAN	FEB	MAR	APR	MAY	JUN	JUL	AUG	SEP	OCT	NOV	DEC
17	017	048	077	108	138	169	199	230	261	291	322	352
18	018	049	078	109	139	170	200	231	262	292	323	353
19	019	950	079	110	140	171	201	232	263	293	324	354
20	020	051	080	111	141	172	202	233	264	294	325	355
21	021	052	081	112	142	173	203	234	265	295	326	356
22	022	053	082	113	143	174	204	235	266	296	327	357
23	023	054	083	114	144	175	205	236	267	297	328	358
24	024	055	084	115	145	176	206	237	268	298	329	359
25	025	056	085	116	146	177	207	238	269	299	330	360
26	026	057	086	117	147	178	208	239	270	300	331	361
27	027	058	087	118	148	179	209	240	271	301	332	362
28	028	059	088	119	149	180	210	241	272	302	333	363
29	029	060	089	120	150	181	211	242	273	303	334	364
30	030		090	121	151	182	212	243	274	304	335	365
31	031		091		152		213	244		305		366

APPENDIX D: APPLICATION FOR SPACE-A AIR TRAVEL (AMC FORM 53)

Below is the "Application for Air Travel", also known as the "Space-A Passenger Booking Card". You or the lead traveler in your group will be required to complete this (or similar) form when you apply in person for Space-A air travel.

(THIS FORM IS SUBJECT TO THE PRIVACY ACT OF 1974 · See Reverse)

PASSENGER'S NAME *(Last, First, Middle Initial)* | GRADE / TITLE | TOTAL SEATS REQUIRED | SPECIAL PAX CODE | TRAVEL ELIGIBILITY | RESERVED | LINE NUMBER

1 2 3 4 5 6 7 8 9 10 11 12 13 14 15 16 17 18 19 20 21 22

AIR MOVEMENT DESIGNATOR | ORIGIN APOE | DESTINATION APOD | PRIORITY | TYPE TRAVEL | SPONSOR | MOVEMENT MONTH | FLIGHT NUMBER | FLIGHT PREFIX | MISSION NUMBER | SUFFIX | FLIGHT CODE

23 24 25 26 27 28 29 30 31 32 33 34 35 36 37 38 39 40 41 42 43

SCHEDULED DEPARTURE TIME | CATEGORY SERVICE CODE | DATE *(Leave express or desired travel date)* | DEPARTURE DATE *(Julian)* | GROUP IDENTIFICATION | NON-REVENUE | SIGN UP DATE *(Julian)* | SIGN UP TIME *(Local – 24 hour clock)* | PASSENGER WEIGHT | BAGGAGE WEIGHT

44 45 46 47 48 49 50 51 52 53 54 55 56 57 58 59 60 61 62 63

SPACE AVAILABLE DESTINATIONS *(Enter up to five destinations desired, use CCs 64–78)* | 1st DEST | 2nd DEST | 3rd DEST | 4th DEST | 5th DEST | RESERVED | TYPE TRANSACTION | PETS: NUMBER | WEIGHT *(Total)*

64 65 66 67 68 69 70 71 72 73 74 75 76 77 78 79 80

SOCIAL SECURITY NUMBER | MCTRANSPORTATION AUTHORIZATION NUMBER | CUSTOMER IDENTIFICATION CODE | DATE LEAVE/PASS STARTS

ACCOMPANYING PASSENGERS

LINE NUMBER	NAME *(Last, First, Middle Initial)*	SPC	RELATION-SHIP	TYPE TRAVEL	PASSENGER IDENTIFICATION *(Social security, passport and visa numbers)*	PASSENGER WEIGHT	BAGGAGE WEIGHT	AGE

AMC FORM SEP 84 53 PREVIOUS EDITIONS ARE OBSOLETE

APPLICATION FOR AIR TRAVEL

APPENDIX E: AUTHENTICATION OF RESERVE STATUS FOR TRAVEL ELIGIBILITY (DD FORM 1853)

Active Duty Status Reserve Component members must present the form below, DD Form 1853, completed and signed by the Reserve organization commander within the previous 180 days when applying for Space-A air travel.

SPACE-A RESERVE STATUS CARD

AUTHENTICATION OF RESERVE STATUS FOR TRAVEL ELIGIBILITY	1. DATE PREPARED (YYMMDD)

PRIVACY ACT STATEMENT

AUTHORITY: 10 USC 8102, 44 USC 3101 and EO 9397.
PRINCIPAL PURPOSE: Use of your SSN is necessary to positively identify you.
ROUTINE USE: Used by Reserve personnel for space available on DoD-owned or controlled aircraft.
DISCLOSURE IS VOLUNTARY: However, failure to disclose it will prevent you from traveling on a DoD-owned or controlled aircraft.

PART A. TO BE COMPLETED BY APPLICANT

2. NAME (Last, First, MI)	3. PAY GRADE	4. BRANCH OF SERVICE	5. SSN

6. UNIT/COMMAND NAME	7. UNIT/COMMAND ADDRESS

I hereby certify that my space-available travel on military aircraft is not for personal gain, or in connection with business enterprise or employment, or to establish a home either overseas or in the United States.

8. SIGNATURE	9. DATE SIGNED (YYMMDD)

PART B. TO BE COMPLETED BY RESERVE ORGANIZATION COMMANDER

The Reservist named above is an active reserve component member and is eligible for space-available transportation on DoD-owned or controlled aircraft in accordance with DoD Regulation 4515.13-R, and is authorized to so travel from ________ to ________
(Not to exceed 30 days.) 10. Date (YYMMDD) 11. Date (YYMMDD)

12. NAME (Last First, MI)	13. PAY GRADE	14. SIGNATURE	15. DATE SIGNED (YYMMDD)

DD Form 1853, 84 APR   Previous editions are obsolete.   U.S. Government Printing Office: [illegible]

APPENDIX F: STANDARD TIME CONVERSION TABLE

The world is divided into 24 time zones or areas. The zero time zone is known as Greenwich Mean Time (GMT) which is physically located at Greenwich, England (UK), on the Meridian of Greenwich. Other areas in this time zone are: Iceland, Ascension Island, England and Scotland. The following table shows major areas and their respective time zones in + and - hours from GMT time. Across the top row of the table each area (zone) to the right of GMT is a plus (+), meaning an hour ahead. Each zone to the left of GMT is a minus (-), meaning an hour behind. The columns down the page are simply the next 24 hours from the base line at the top. For example, if you are in Germany, Italy or Spain (GMT +1), the local time is 0800 hours, and you wish to know the time in San Francisco, CA, USA, you read down the GMT +1 column to 0800 hours and left to GMT -8 (Pacific Time USA, San Francisco, CA, USA) where it is 2300 hours. Practice with these tables until you are proficient.

Note: This chart is for planning purposes only, as local times may vary due to local conditions, such as daylight savings time, etc. For exact local times, consult the DoD Foreign Clearance Guide, DoD Flight Information Publication, or Air Almanac.

- NOTES -

APPENDIX F, Continued

STANDARD TIME CONVERSION TABLE:

-12	-11	-10	-9	-8	-7	-6	-5	-4	-1	GMT	+1
0600	0700	0800	0900	1000	1100	1200	1300	1400	1700	1800	1900
0700	0800	0900	1000	1100	1200	1300	1400	1500	1800	1900	2000
0800	0900	1000	1100	1200	1300	1400	1500	1600	1900	2000	2100
0900	1000	1100	1200	1300	1400	1500	1600	1700	2000	2100	2200
1000	1100	1200	1300	1400	1500	1600	1700	1800	2100	2200	2300
1100	1200	1300	1400	1500	1600	1700	1800	1900	2200	2300	2400
1200	1300	1400	1500	1600	1700	1800	1900	2000	2300	2400	0100
1300	1400	1500	1600	1700	1800	1900	2000	2100	2400	0100	0200
1400	1500	1600	1700	1800	1900	2000	2100	2200	0100	0200	0300
1500	1600	1700	1800	1900	2000	2100	2200	2300	0200	0300	0400
1600	1700	1800	1900	2000	2100	2200	2300	2400	0300	0400	0500
1700	1800	1900	2000	2100	2200	2300	2400	0100	0400	0500	0600
1600	1900	2000	2100	2200	2300	2400	0100	0200	0500	0600	0700
1900	2000	2100	2200	2300	2400	0100	0200	0300	0600	0700	0800
2000	2100	2200	2300	2400	0100	0200	0300	0400	0700	0800	0900
2100	2200	2300	2400	0100	0200	0300	0400	0500	0800	0600	1000
2200	2300	2400	0100	0200	0300	0400	0500	0600	0900	1000	1100

APPENDIX F, Continued

-12	-11	-10	-9	-8	-7	-6	-5	-4	-1	GMT	+1
2300	2400	0100	0200	0300	0400	0500	0600	0700	1000	1100	1200
2400	0100	0200	0300	0400	0500	0600	0700	0800	1100	1200	1300
0100	0200	0300	0400	0500	0600	0700	0800	0900	1200	1300	1400
0200	0300	0400	0500	0600	0700	0800	0900	1000	1300	1400	1500
0300	0400	0500	0600	0700	0800	0900	1000	1100	1400	1500	1600
0400	0500	0600	0700	0800	0900	1000	1100	1200	1500	1600	1700
0500	0600	0700	0800	0900	1000	1100	1200	1300	1600	1700	1800

Zone GMT -12: Kwajalein Atoll. -11: Midway Island, Pago Pago, Canton. -10: HawaII, Earickson. -9: EImendorf. -8: US Pacific Time. -7: US Mountain Time. -6: US Central Time. -5: US Eastern Time, Panama, Cuba, -4: Bermuda, Puerto Rico, Greenland. -1: Azores. GMT: Iceland, Ascension Islands, England, Scotland. +1: Germany, Italy, Spain.

- NOTES -

APPENDIX F, Continued

+2	+3	+6	+7	+8	+9	+9:30	+10	+12
2000	2100	2330	0100	0200	0300	0330	0400	0600
2100	2200	0030	0200	0300	0400	0430	0500	0700
2200	2300	0130	0300	0400	0500	0530	0600	0800
2300	2400	0230	0400	0500	0600	0630	0700	0900
2400	0100	0330	0500	0600	0700	0730	0800	1000
0100	0200	0430	0600	0700	0800	0830	0900	1100
0200	0300	0530	0700	0800	0900	0930	1000	1200
0300	0400	0630	0800	0900	1000	1030	1100	1300
0400	0500	0730	0900	1000	1100	1130	1200	1400
0500	0600	0830	1000	1100	1200	1230	1300	1500
0600	0700	0930	1100	1200	1300	1330	1400	1600
0700	0800	1030	1200	1300	1400	1430	1500	1700
0800	0900	1130	1300	1400	1500	1530	1600	1800
0900	1000	1230	1400	1500	1600	1630	1700	1900
1000	1100	1330	1500	1600	1700	1730	1800	2000
1100	1200	1430	1600	1700	1800	1830	1900	2100
1200	1300	1530	1700	1800	1900	1930	2000	2200
1300	1400	1630	1800	1900	2000	2030	2100	2300
1400	1500	1730	1900	2000	2100	2130	2200	2400
1500	1600	1830	2000	2100	2200	2230	2300	0100
1600	1700	1930	2100	2200	2300	2330	2400	0200
1700	1800	2030	2200	2300	2400	0030	0100	0300
1800	1900	2130	2300	2400	0100	0130	0200	0400
1900	2000	2230	2400	0100	0200	0230	0300	0500

Zone GMT +2: Greece, Egypt. +3: Dhahran, Turkey, Bahrain. +6: Diego Garcia. +7: Thailand. +8: Philippines, Taiwan: Okinawa JA, Korea. +9:30: Alice Springs AU, Woomera AU. +10: Guam, Richmond AU. +12: Wake Island, New Zealand.

APPENDIX G: BOARDING PASS/TICKET RECEIPT (AMC FORM 148/2)

This or similar form will be used to record boarding, baggage, meals, and other charges.

AMC- **BOARDING PASS/TICKET/RECEIPT**

NAME (*Last, First, Middle*)	FLIGHT NO.	GATE	BOARDING TIME	SEAT NO.	SMOKING
DESTINATION	DEPARTURE DATE	BAGGAGE WEIGHT/PIECES	EXCESS WEIGHT	MEAL COST $	
VIA	OTHER WEIGHT	MEAL (*Kind/Type/Quantity*)	BAGGAGE COST		
ORIGIN	REMARKS	OTHER COST			
CARRIER	REASON/DATE OF REFUND	CASH COLLECTED $			
PASSENGER SERVICE AGENT SIGNATURE	PASSENGER SIGNATURE				

AMC Form 148/2, OCT 87 — REPLACES AMC Form 124a, AUG 79 WHICH WILL BE USED UNTIL STOCK IS EXHAUSTED. — PASSENGER COPY

APPENDIX H: INTERNATIONAL CERTIFICATES OF VACCINATION AND PERSONAL HEALTH HISTORY (PHS FORM 731)

This document provides for the recording of international certificates of vaccination and revaccination in both the English and French languages, and the personal health history of international travelers. This document, with current health entries, is required as a Personnel Entrance Requirement for many foreign countries.

I. INTERNATIONAL CERTIFICATES OF VACCINATION

AS APPROVED BY

THE WORLD HEALTH ORGANIZATION

(EXCEPT FOR ADDRESS OF VACCINATOR)

CERTIFICATS INTERNATIONAUX DE VACCINATION APPROUVES
PAR L'ORGANISATION MONDIALE DE LA SANTE

(SAUF L'ADRESSE DU VACCINATEUR)

II. PERSONAL HEALTH HISTORY

TRAVELERS NAME Nom du voyageur

ADDRESS
ADRESSE (Number——Numéro) (Street—Rue)

(City--Ville)

(Country--Départment) (State--État)

U.S. DEPARTMENT OF HEALTH, EDUCATION, AND WELFARE
PUBLIC HEALTH SERVICE

READ INSTRUCTIONS CAREFULLY

PHS--731
Rev. 9-66

APPENDIX H, continued: INTERNATIONAL CERTIFICATES OF VACCINATION AND PERSONAL HEALTH HISTORY (PHS FORM 731)

INTERNATIONAL CERTIFICATE OF VACCINATION OR REVACCINATION AGAINST SMALLPOX
CERTIFICAT INTERNATIONAL DE VACCINATION OU DE REVACCINATION CONTRE LA VARIOLE

This is to certify that ... sex
Je soussigné(e) certifie que .... sexe

whose signature follows ... date of birth
dant la signature suit .... né(e) le

has on the date indicated been vaccinated or revaccinated against smallpox with a freeze-dried liquid vaccine certified to fulfill the recommended requirements of the World Health OrganIzatIon. a été vacciné(e) ou revacciné(e) contre la variole á la date indiqueé ci-dessous avec un vaccin Iyophilisé ou liquide certifié conforme aux normes recommandée par L´ Organisation mondiale de Ia Santé.

INTERNATIONAL CERTIFICATE OF VACCINATION OR REVACCINATION AGAINST YELLOW FEVER
CERTIFICAT INTERNATIONAL DE VACCINATION OU DE REVACCINATION CONTRE LA FIEVRE JUANE

This is to certify that ... sex
Je soussigné(e) certifie que .... sexe

whose signature follows ... date of birth
dant la signature suit .... né(e) ie

has on the date indicated been vaccinated or revaccinated against yellow fever.
a été vacciné(e) ou revacciné (e) contre la fievre juane a la date indiquée.

INTERNATIONAL CERTIFICATE OF VACCINATION OR REVACCINATION AGAINST CHOLERA
CERTIFICATE INTERNATIONAL VACCINATION OU DE REVACCINATION CONTRE LE CHOLERA

This is to certify that ... sex
Je soussigné(e) certifie que .... sexe

whose signature follows ... date of birth
dant la signature suit .... né(e) le

has on the date indicated been vaccinated or revaccinated against cholera.
a été vacciné(e) ou revacciné(e) contre le choléra á date indiquée.

APPENDIX I: BAGGAGE IDENTIFICATION (DD FORM 1839, AMC FORM 20-ID, AND USAF FORM 94)

These baggage identification tags (and others) are used to identify checked and cabin luggage.

BAGGAGE IDENTIFICATION

NAME *(Last, First, M.I.)*

STREET ADDRESS *(Home or Unit/APO)*

CITY, STATE AND ZIP CODE

DD FORM 80 SEP 1839 USE PREVIOUS EDITION.

SOCIAL SECURITY NUMBER

TELEPHONE NO. (Home/Work) (Area Code is needed)

NAME (Last, First, M.I.)

ADDRESS (Home or Unit/APO) CITY, STATE & ZIP CODE

MAC FORM 20-ID, AUG 84

SUU

TRAVIS AFB, CALIFORNIA

MISSION NUMBER/DATE

FROM

000000

Strap Check - Not a Claim Check

Checked Baggage
USAF FORM 94

APPENDIX J: STATE, POSSESSION, AND COUNTRY ABBREVIATIONS

STATE ABBREVIATIONS

AL-Alabama
AK-Alaska
AR-Arkansas
AZ-Arizona
CA-California
CO-Colorado
CT-Connecticut
DC-District of Columbia
DE-Delaware
FL-Florida
GA-Georgia
HI-Hawaii
IA-Iowa
ID-Idaho
IL-Illinois
IN-Indiana
KS-Kansas
KY-Kentucky
LA-Louisiana
MA-Massachusetts
MD-Maryland
ME-Maine
MI-Michigan
MN-Minnesota
MO-Missouri
MS-Mississippi
MT-Montana
NE-Nebraska
NC-North Carolina
ND-North Dakota
NV-Nevada
NH-New Hampshire
NJ-New Jersey
NM-New Mexico
NY-New York
OH-Ohio
0K-Oklahoma
OR-Oregon
PA-Pennsylvania
RI-Rhode island
SC-South Carolina
SD-South Dakota
TN-Tennessee
TX-Texas
UT-Utah
VA-Virginia
VT-Vermont
WA-Washington
WI-Wisconsin
WV-West Virginia
WY-Wyoming

POSSESSION ABBREVIATIONS

AS-American Samoa
GU-Guam
JO-Johnston Atoll
KA-Kwajalein Atoll
MI-Marshall Islands
MP-Northern Mariana Island
MW-Midway Island
PW-Palau (TTPI)
PR-Puerto Rico
TTPI-Trust Territory of Pacific Islands
VI-U.S. Virgin Islands
WK-Wake Island

*COUNTRY ABBREVIATIONS

AG-Argentina
AI-Ascension Island (UK)
AN-Antigua/Barbuda
AU-Australia
BA-Bahrain
BB-Barbados
BE-Belgium
BH-Bahamas
BM-Bermuda (UK)
BO-Bolivia
BR-Brazil
BZ-Belize
CD-Chad
CH-Chile
CL-Columbia
CN-Canada
CR-Crete (GR)
CS-Costa Rica
CU-Cuba
CY-Cyprus
DN-Denmark
DR-Dominican Republic
EC-Ecuador
EG-Egypt
ES-El Salvador
FM-Federated States of Micronesia
GE-Germany
GL-Greenland (DN)
GR-Greece
GT-Guatemala
GY-Guyana
HA-Haiti
HO- Honduras
HK-Hong Kong (UK)
IC-Iceland
IE-Indonesia
IO-Indian Ocean (Diego Garcia) (UK)
IR-Ireland
IS-Israel
IT-Italy
JA-Japan
JR-Jordan
KE-Kenya
LI-Liberia
MA-Malaysia
MX-Mexico
NG-Niger
NI-Nicaragua
NO-Norway
NT-Netherlands
NZ-New Zealand
OM-Oman
PE-Peru
PG-Paraguay
PN-Republic of Panama
PO-Portugal (Azores)
RK-Republic of Korea
RP-Republic of the Philippines
SA-Saudi Arabia
SE-Senegal
SG-Singapore
SI-Solomon Islands
SM-Somalia
SP-Spain
SR-Suriname
SU-Sudan
TH-Thailand
TU-Turkey
UG-Uruguay
UK-United Kingdom
US-United States
VE-Venezuela
ZA-Zaire

*Countries for which abbreviations are provided have been limited to those countries where Space-A travel is generally scheduled. Prime governing nations are noted within parentheses.

APPENDIX K: AIR PASSENGER COMMENTS (AMC FORM 253)

As a Space-A passenger you are encouraged to use this (or similar) form to report positive and negative information to managers of the system who are in a position to correct deficiencies and/or reward outstanding performance of duty. (Form size adjusted to fit on this page. Actual size is larger) *Please send a courtesey copy of your comments to : Military Living's R&R Space-A Report®, P.O. Box 2347, Falls Church, VA 22042-0347.*

AIR PASSENGER COMMENTS

Please provide a copy to terminal management by placing in the slot marked for Squadron/Port Operations Officer. Terminal addresses are listed on the reverse in case you desire to mail your comments to a terminal you have passed through. Your comments to Squadron/Port Operations Officers will let them take immediate action. If you feel we need to know about a particular item, send a copy of your comments to us: HQ Air Mobility Command
Aerial Port Operations Division (AMC/XON)
402 Scott Drive, Room 132
Scott AFB, IL 62225-5363

AIR MOBILITY COMMAND

COMMENTS

To assist us please provide the following information when applicable.

NAME *(Last, First, M.I.) (Optional)*	GRADE *(Optional)*	DUTY ADDRESS *(Optional)*	DUTY PHONE *(Optional)*

FLIGHT NUMBER	DEPARTING FROM	DESTINATION	DATE FORM PREPARED *(Day, Month, Year)*

AMC FORM 253, MAY 93 REPLACES AMC FORM 253, JUN 92 WHICH IS OBSOLETE

APPENDIX L: A BRIEF DESCRIPTION OF AIRCRAFT ON WHICH MOST SPACE-A TRAVEL OCCURS

The following transport, tanker, and special mission aircraft are used by the military services (USPHS and NOOA do not have aircraft which are suitable for Space-A travel) for missions having Space-A air opportunities. **Only the major channel and support aircraft are listed.** We have not listed minor, some special mission, and helicopter (rotary wing) aircraft due to space limitations. **We have provided for you a brief description of each aircraft with emphasis on performance and passenger accommodations.** The total number of each aircraft changes in the inventory due to acquisitions, conversions, reconfiguration, and attrition. **Our best estimate of current specific aircraft inventories are listed below.**

C-5A/B GALAXY

The C-5A/B is a long-range, air-refuelable, heavy logistics transport which is capable of airlifting loads up to 291,000 pounds. This aircraft was developed, designed and configured to meet a wide range of military airlift missions. This is the "Free World's" largest aircraft.

PROGRAM/PROJECT CONTRACTOR: Lockheed Aeronautical Systems Company.
POWER SOURCE: Four General Electric TF39-GE-1C turbofan engines.
Each engine has 43,000 lbs of thrust.
DIMENSIONS: Wing span is 222 ft, 8.5 in. Length is 247 ft, 10 in. Height is 65 ft, 1.5 in.

WEIGHTS: Empty weight is 374,000 lbs. Maximum payload is 261,000 lbs. Gross weight is 837,000 lbs.
PERFORMANCE: Maximum speed at 25,000 ft is 571 mph. Service ceiling with 615,000 lbs gross weight is 35, 750 ft. Range with maximum payload is 3,434 miles and range with maximum fuel is 6,469 miles. Between 1982-1987 the 77 C-5A's were upgraded to C-5B capabilities. From 1985-1989, 50 C5B's were acquired.
FACILITIES: Aircraft crew of six. Relief crew/rest area of 15. **Seating for 75 passengers, 2nd deck airline type seats facing to the rear of the aircraft for safety purposes.** cargo, 1st deck, 36 standard 463L pallets or mounted weapons and vehicles or a maximum of 340 passengers in a wide-body jet configuration. There is a program to repaint all USAF C-5A/B's flat grey. AMC has control of all C-5A/B's.
INVENTORY: Total USAF 127.

C-009A/E NIGHTINGALE

This aircraft was designed as a commercial airliner. The DC-9 Series 30 commercial aircraft was reconfigured, modified and equipped to perform aeromedical (air ambulance) airlift transport missions. The C-009A/C performs aeromedical missions in CONUS, and in the European and Pacific Theaters.

PROGRAM/PROJECT CONTRACTOR: Douglas Aircraft Company. Division of McDonnell Douglas Corporation.
POWER SOURCE: Two Pratt & Whitney JT8D-9 turbofan engines. Each engine produces 14,500 lbs of thrust.
DIMENSIONS: Wing span is 93 ft, 3 in. Length is 119 ft, 3 in. Height is 27 ft, 6 in.
WEIGHT: Gross weight 108,000 lbs.

PERFORMANCE: The maximum cruising speed at 25,000 ft is 565 mph. Ceiling is 35,000 ft. Range is in excess of 2,000 miles.
FACILITIES: Aircraft crew of three (includes flight mechanic and spare parts) and five medical staff. There can be a combination of 40 litter (stretcher) or 40 ambulatory patients. Most MEDEVAC patients are ambulatory, that is, they can walk but may be put in a litter for comfort. The ambulatory seats are spacious airline type seats. These are the seats used by Space-A passengers.
INVENTORY: Twenty-one in CONUS, four in Europe, three in Pacific, for a total inventory of 28 aircraft configured for aeromedical missions. Three are specifically configured C-9C's which are assigned for Presidential and related missions. **The USN has 29 each C-9B SKYTRAIN II aircraft procured in FY 1985 to meet major Navy logistics requirements. This aircraft is configured for cargo and passenger (airline type seats, up to approximately 100). Total 60.**

C-17A GLOBEMASTER III

This is a new aircraft which is now undergoing initial operational testing. It is a heavy-lift, air-refuelable, cargo transport designed to meet inter-theater and intra-theater airlift for all types of cargo and passengers. This aircraft will be capable of using unimproved landing facilities (runways - 90 ft wide x 3,000 ft long). The initial operational capability (IOC) date is scheduled for FY 1994. A total of 40 aircraft have been funded through FY 1995. The planned total acquisition is 120 aircraft. **The passenger configurations for this aircraft have not been established.**

PROGRAM/PROJECT CONTRACTOR: McDonnell Douglas Aerospace Transport Aircraft Division of McDonnell Douglas Aerospace.
POWER SOURCE: Four Prat & Whitney F117-PW 100 turbofans; each 40,000 lbs of thrust on each aircraft.
DIMENSIONS: Wing span is 169 ft 10 in. Lenght is 174 ft. Height is 55ft 1 in.

WEIGHT: Payload 172,000 lbs, Gross weight 585,000 lbs.
PERFORMANCE: Cruising speed (estimated) 518 mph, range with 160,000 lbs payload is 2,765 miles.
FACILITIES: The passenger and cargo configurations for this aircraft have not been determined.
INVENTORY: Total USAF 20.

C-21A EXECUTIVE AIRCRAFT

There is a group of executive type aircraft in use in all of the Military Services. The C-21A is typical of these aircraft. **We will list the data for the C-21A and then list the inventory and passenger capacity of executive type aircraft in the Military Services.**

PROGRAM/PROJECT CONTRACTOR: Learjet Corporation.
POWER SOURCE: Two Garrett TFE731-2-turbojet engines. Each engine has 3,500 lbs thrust.
DIMENSIONS: Wing span is 39 ft, 6 in. Length is 48 ft, 8 in. Height is 12 ft, 3 in.
WEIGHT: Gross 18,300 lbs.
PERFORMANCE: Cruising speed is Mach 0.81. Service ceiling is 45,000 ft. Range with maximum passengers is 2,420 miles and with maximum cargo load is 1,653 miles.
FACILITIES: Aircraft crew of two. **Eight passengers in airline type seats,** or cargo of 3,153 lbs. Also convertible to aeromedical (MEDEVAC) configuration.
INVENTORY: Total 498.

C-12A-J HURON (8 Passengers (PAXs): USAF-77,

HU-25A GUARDIAN (APPROX 12 PAXs): USCG-41,

C-20A/B GULFSTREAM III/IV (14-18 PAXs): USAF-22

C-21A EXECUTIVE AIRCRAFT (8 PAXs): USAF-88,

C-22B (BOEING 727 (APPROX 100 PAXs): USAF-4,

C-23A SHERPA (APPROX 8 PAXs): USAF-13,

VC-25A Presidential Transport, USAF-2, (53 ON ORDER), C-27A s

C-26-A FAIRCHILD METRO III (19-20 PAXs): USAF -13

(53 ON ORDER), C-27A STOL (53 PAXs): USAF-5,

C-29A (125-800 BUSINESS JET, APPROX 8 PAXs): USAF-6.

The U.S. Army operates a fleet of C-12, U-21 and older aircraft of approximately 200 in number. Each aircraft can seat approximately 8 passengers.

C130A-H HERCULES

The C-130 Hercules is a very versatile aircraft which is used to perform a wide range of missions for all of the military services. The aircraft has been used mainly in a cargo and passenger role. It has also been used in specialized combat, electronic warfare, Arctic ice cap resupply, aerial spray, aeromedical MEDEVAC, and aerial refueling among many similar missions. This aircraft is found in the inventory of all the military (Armed) services.

PROGRAM/PROJECT CONTRACTOR: Lockhead Aeornautical Systems Company.
POWER SOURCE: Four Allison T-56-A-15 turboprop engines. Each engine has 4,508 ehp.
DIMENSIONS: Wing span is 132 ft, 7 in. Length is 97 ft, 9 in. Height is 38 ft, 3 in.

PERFORMANCE: The maximum cruising speed at 20,000 ft is 374 mph. The service ceiling for 130,000 lbs is 33,000 ft. The range with maximum payload is 2,356 miles.
FACILITIES: Aircraft crew of five, **92 passengers in commercial airline type seats,** 74 litter patients, five 463L standard pallets, and assorted mounted weapons and vehicles. Seating ranges from side "bucket" seats along the sides of the aircraft to airline type seating with aisles and facing to the rear. The noise level is extremely high in this aircraft. Ear plugs are highly recommended for all passengers and crew.
INVENTORY: Total 1,271. C-130A-H and HC-130H/N/P: USAF-APPROX 1,100. USN-97, USMC-42 (KC-130), USCG-30 (HC-130), USA-1 (EW MISSIONS).

KC-135A-R STRATOTANKER

This stratotanker was designed to military specifications. The aircraft is similar in size and design appearance to the commercial 707 aircraft but there the similarity ends. The KC-135 has different internal structural designs and materials which stress the ability to operate at high gross weights. The fuel carried in this tanker is located in the "wet wings" and in the fuel tanks below the floor in the fuselage. Passengers traveling on this aircraft are allowed, subject to mission restraints, to observe the Air to Air Refueling Operations which usually take place over the world's oceans.

PROGRAM/PROJECT CONTRACTOR: Boeing Military Airplanes.
POWER SOURCE: Four CFM international F108-CF-100 turbofan engines. Each engine has 22,224 lbs of thrust.**DIMENSIONS:** Wing span is 130 ft, 10 in. Length is 136 ft, 3 in. Height is 38 ft, 4 in.
WEIGHT: Empty weight is 119,231 lbs. Gross 322,500 lbs.
PERFORMANCE: The maximum speed at 30,000 ft is 610 mph. Service ceiling 50,000 ft. Range with 12,000 lbs of transfer fuel is 11,192 miles.
FACILITIES: Aircraft crew of 4 or 5. **Maximum of 80 passengers in airline type seats facing to the rear of the aircraft.**
INVENTORY: USAF 552.

C-135B STRATOLIFTER

This aircraft is similar to the KC-135 Stratotanker without the refueling equipment. These aircraft were initially purchased as an interim cargo/passenger aircraft placed in service before delivery of the C-141's. The appearance of this aircraft is similar to the KC-135.

PROGRAM/PROJECT CONTRACTOR: Boeing Military Airplanes.
POWER SOURCE: Four CFM international F108-CF-100 turbofan engines. Each engine has 22,224 lbs of thrust.
DIMENSIONS: Wing span is 130 ft, 10 in. Length is 134 ft, 6 in.
Height is 38 ft, 4 in.
WEIGHT: Empty 102,300 lbs. Gross 275,000 lbs.
PERFORMANCE: Maximum speed 600 mph. Range with 54,000 lb payload is 4,625 miles.
INVENTORY: USAF 48

VC-137B/C STRATOLINER

This is a special mission aircraft which has been modified from the commercial Boeing 707 transport. Two of these aircraft were the original "Air Force One" aircraft used by past United States Presidents.

PROGRAM/PROJECT CONTRACTOR: The Boeing Company.
POWER SOURCE: Four Pratt & Whitney JT3D-3 turbofan engines. Each engine has a 17,200 lb thrust.
DIMENSIONS: VC-137B: Wing span is 130 ft, 10 in. Length 144 ft, 6 in. Height 42 ft, 10 in. VC137-C: Wing span is 145 ft, 9 in. Length is 152 ft, 11 in. Height is 42 ft, 5 in.
WEIGHT: VC-137B: Gross 258,000 lbs. VC-137C: Gross 322,000 lbs.
PERFORMANCE: VC-137C: Maximum speed 627 mph. Service ceiling 42,000 ft. Range 5,150 miles.
FACILITIES: This is a special mission aircraft with a variety of configurations. There are full-service galleys, dining, sleeping berths, and airline type seating.
INVENTORY: USAF 7.

C-141A/B STARLIFTER

The C-141A/B STARLIFTER transport has undergone extensive modification to extend the airframe and modernization to all aspects of the aircraft. The result is a modern air transport which is fully capable of performing many missions from routine cargo and passengers to inter-theater MEDEVAC and humanitarian missions around the world. All of the C-141A/B fleet are scheduled for repainting to a flat grey.

POWER SOURCE: Four Pratt & Whitney TF33-P-7 turbofan engines. Each engine has 21,000 lbs of thrust.
DIMENSIONS: Wing span is 159 ft, 11 in. Length is 168 ft, 3.5 in. Height is 39 ft, 3 in.
WEIGHT: operating 149,000 lbs. Maximum payload 89,000 lbs. Gross 343,000 lbs.
PERFORMANCE: Maximum cruising speed is 566 mph. Range with maximum payload is 2,293 miles without air refueling.
FACILITIES: Air crew of five. **200 passengers in commercial airline seats facing to the rear of the aircraft.** 103 litter patients plus attendants. Cargo on 13 standard 463L pallets or alternate mounted weapons, vehicles or other cargo.
INVENTORY: USAF 270.

KC-10A EXTENDER

This advanced tanker/cargo aircraft is based on the commercial DC-10 Series, 30 CF. It has been modified to include fuselage fuel cells, aerial refueling operator station and boom. Military avionics have been added. The aircraft is fit to perform a role of extending and enhancing worldwide military mobility. The latest modifications to this aircraft are wing-mounted air-refueling pods designed to supplement the basic system and increase capability.

PROGRAM/PROJECT CONTRACTOR: Douglas Aircraft Company, Division of McDonnell Douglas Corporation.
POWER SOURCE: Three General Electric CF-6-50C2 turbofan engines. Each engine has 52,500 lbs of thrust.

DIMENSIONS: Wing span is 165 ft, 4.5 in. Length is 181 ft, 7 in. Height is 58 ft, 1 in.
WEIGHT: Gross 590,000 lbs.

PERFORMANCE: Cruising speed Mach 0.825. Service ceiling 42,000 ft range with maximum cargo 4,370 miles.
FACILITIES: Aircraft crew of four. **75 passengers in commercial airline seats facing to the rear of the aircraft.** 27 standard 463L pallets. Maximum cargo payload 169,409 lbs.
INVENTORY: USAF 59.

P-3C-ORION

This is a propeller-driven aircraft which has been used by the U.S. Navy since 1958 in an Anti-Submarine Warfare (ASW) role. Many improvements have been incorporated in the basic airframe over the years. The latest improvements allow the aircraft to detect, track and attack quieter new generation submarines. **The replacement P-7A program with Lockheed as the contract was terminated in July 1990. The USN is investigating alternative programs.**

PROGRAM/PROJECT CONTRACTOR: Lockheed.
POWER SOURCE: Four Allison T-56-A-14 turboprop engines. Each engine has 4,900 ehp.
DIMENSIONS: Wing span is 100 ft. Length is 117 ft. height is 34 ft.
WEIGHT: Gross wieght is 139,760 lbs.

PERFORMANCE: Maximum speed 473 mph. Cruise speed 377 mph. Ceiling 28,300 ft.
FACILITIES: Aircraft crew of 10. **18 passengers in airline seats.**
INVENTORY: USN-133. Seventy-three older aircraft are to be retired in the very near future, thus reducing the number of aircraft in regular and reserve P-3 squadrons. **Total-247.**

-NOTES-

APPENDIX M: SPACE-A QUESTIONS AND ANSWERS

One of the biggest fringe benefits, dollar-wise, for uniformed services personnel and their family members is Space-A air travel on U.S. military owned and operated aircraft. While there are some old pros who know all the ropes, having learned the hard way by flying Space-A, there are those who are a bit afraid to jump into the unknown. **This appendix is especially for those who want to know as much as they can about Space-A air travel.** Answers are based on information available to us at press time. Because policies can change or be interpreted differently, **these general answers must be regarded only as guides - not rules.** Specific questions, particularly those dealing with changes in policy, should be directed to military officials who are the final authority on the subject. We have divided the questions and answers into general functional categories. We hope that this will aid readers in locating questions and answers in which they have a special interest.

BAGGAGE

01. How much baggage can Space-A passengers check? Each Space-A passenger (regardless of age) can check two pieces of baggage totaling 140 pounds. Air Mobility Command (AMC) limits the size of each item to 62 linear inches. This measurement is obtained by adding together the item's length, width, and height. The rules permit some exceptions to the 62 linear inches size limitation. All duffel bags, sea bags, Air Force issue B-4 bags, and civilian-origin versions that have the same approximate dimensions can be checked. Similarly, the size restrictions do not apply to golf bags with golf clubs, snow skis, folding bicycles, fishing equipment, musical instruments and rucksacks. Any one of these oversized items listed above may be checked if it is the only piece checked and meets weight requirements of 140 pounds total.

02. We have heard that families and other groups can "pool" their baggage authorization. What's the story? Space-A passengers traveling together as a group - that is, listed on a single Military Transportation Authorization or AMC Form 53 (Application for Air Travel) - may pool their baggage authorization, so long as the total number of checked pieces does not exceed the number of travelers times 140 pounds, i.e. a five person family travel group could not exceed 700 pounds (5 x 140 pounds = 700 pounds).

03. How much baggage can I carry with me into the passenger cabins? All passengers boarding the aircraft can carry on one or more pieces so long as they fit under the passenger's seat, in the overhead compartment, or other approved storage area, e.g., closets for hang-up garment bags. If available storage space is important to your baggage carrying needs, inquire at the terminal regarding storage areas for carry-on baggage before checking your baggage for a particular flight. As a guideline carry-on bags should not exceed 45 linear inches (length + width + height = 45 inches). **Passengers traveling with babies can also carry on any Federal Aviation Administration (FAA) approved infant car seat regardless of any other baggage.** Each AMC facility has a list of the FAA approved car seats. Passengers can call the FAA at Tel: (202) 426-3800 to determine if new seats have been added to the list of approved seats.

04. Is the baggage limit the same for all aircraft? No. The baggage limit for smaller executive aircraft and the C-009A/C Nightingale is considerably less. On small two-engine executive and operational support aircraft, **the baggage limit for Space-A passengers is 30 pounds.** Also, on the C-009A/C aircraft the size limit for carry-on baggage is 18" long, 5" wide and 19" high or 42" overall.

05. As a Space-A passenger, may I pay for excess checked baggage over 140 pound or two pieces? No. Only duty status passengers may pay for excess baggage.

Note see Appendix N, SPACE-A TRAVEL TIPS, for more information on baggage.

ELIGIBILITY

06. May all Active Duty and Retired members of all the Uniformed Services fly Space-A? Yes. All Active Duty and Retired members (as well as their eligible family members) of all seven uniformed services: U.S. Army, U.S. Navy. U.S. Marine Corps, U.S. Coast Guard, U.S. Air Force, U.S. Public Health Service Officer Corps, and National Oceanic and Atmospheric Administration Officer Corps may fly Space-A as provided for in DoDD 4515.13-R as revised. Dependent family members may only accompany their sponsor on flights going overseas and in overseas areas.

07. May National Guard and Reservists fly Space-A? National Guard members and Reservists, in an Active paid status, may fly anywhere in CONUS, Alaska, Hawaii, Puerto Rico, Guam, American Samoa, and the U.S. Virgin Islands. Guard and Reserve members cannot fly Space-A to a foreign country. Guard and Reserve members must have the ID Card, DD Form 2 (Red), and DD Form 1853, Authentication of Reserve Status for Travel Eligibility (authenticated by the Unit Commander within the last 6 months). The same is true of Guard and Reserve personnel who have received official notification of retirement eligibility but have not reached retirement age (60). This "Gray Area" retirement eligible group must present their ID cards (Red) and retirement eligibility notices (letters).

08. When may National Guard and Reservists eligible family members fly Space-A? When the sponsor retirees and receives retired pay and full benefits at age 60, eligible family members may then fly Space-A. Family members must be accompanied by their sponsor when flying Space-A, and may only fly on flights going overseas and in the overseas area.

09. Is there any difference in Space-A rules regarding eligibility for Active Duty versus Retired service members? Yes. First of all, Active Duty personnel have priority (Categories 2 and 3) on Space-A flights at all times. Other differences include the fact that Active Duty personnel may take their "dependent" mothers and fathers (who have ID Cards DD Form 1173), with them on Space-A trips. **Dependent in-laws are NOT included in this privilege.** Retired members do not have this privilege, and Retired members and their families travel in Category 4.

10. I am a 100% disabled American veteran (DAV). I've heard that some of us can fly Space-A and some can't. Could you give me more information on 100% DAV's and Space-A? Disabled American veterans must be RETIRED from a uniformed service to qualify for Space-A travel. Those members who were separated in lieu of being retired are not eligible. Here's an easy way to check your eligibility. If your monthly retired check is paid by a uniformed services finance center, e.g. U.S. Army, and your ID card is DD Form 2 (old cards are gray in color; new cards are blue), you can fly Space-A. If you are paid by the Veterans Administration and your ID card is a Form 1173 (butterscotch in color), you cannot fly Space-A. The color of ID cards and their form numbers are the key to being allowed to sign-up for a Space-A flight. The DD Form 1173 is the same ID form used by dependents. In any case, dependents are not generally allowed to fly Space-A without their sponsors, so this butterscotch color card is a red flag alerting the officials at the Space-A desk that the carrier of the DD Form 1173 is not eligible to fly Space-A unaccompanied.

11. Who may fly on National Guard and Reserve flights of the Military Services? All uniformed services personnel and their eligible dependents may fly on most National Guard and Reserve flights, depending upon the mission. The National Guard and Reserve have some of the best flights available. The catch is that they are not generally scheduled flights. Many different types of flight missions are given to National Guard and Reserve units, therefore, one can often find some very special flights to places not normally seen on flight schedules. Most National Guard and Reserve departure locations are listed in **Military Living's Military Space-A Air Opportunities Around The World** book.

12. Are Active Duty personnel in a leave or pass status, traveling Space-A, always required to wear the service uniform? No. All Active Duty members (excluding USMC personnel) in a leave or pass status and traveling Space-A on military department owned and operated aircraft are now not required to wear the class A or B uniform of their service.

"You may think it's absolutely ingenious. I think it detracts from the uniform."

13. May an Active Duty service member use Space-A to take dependents to his/her unaccompanied duty station overseas or back from overseas to CONUS after the unaccompanied duty tour is completed? No. Family members may use Space-A only when they are with the sponsor on an accompanied tour (on service orders) overseas. The Space-A privilege is intended only for a visit to an overseas or CONUS area on a round-trip basis with the sponsor. **Space-A cannot be used to establish a home for dependents overseas or in CONUS.**

14. May an Active Duty service member sign-out on leave, sign-up (register) for Space-A and if there is a wait for the flight, go back to work to avoid loss of leave time? When registering for Space-A travel, either by FAX, mail/courrier or in person, the member must have an approved leave or pass authorization effective on or before the date of registration. You must show your approved leave with an effective date on or before your sign-up date. If a member registers for Space-A travel but voluntarily returns to work during the intervening days before the actual flight departure, leave will be charged for those days. **You must be on leave throughout your entire Space-A leave travel period.**

"Hey dad, there's one of those new Korean fighters with a red star on it."

15. What does it mean to be "bumped"? The mission needs of space required passengers or cargo may require the removal of Space-A passengers at any point. If removed after being manifested on a flight or en route, you may re-register with the date and time adjusted to reflect the date and time of registration at the point of origin. The Space-A passengers will be placed no higher than the bottom of their Category on the Space-A register. **Space-A passengers cannot be "bumped" by other Space-A passengers.**

16. What can service families do if they become extremely ill while overseas and need to return to the United States? Air medical evacuation (MEDEVAC) through AMC is available to Active Duty, Retired and their eligible family members. Space-A travelers should get in touch with a U.S. military medical facility, preferably a hospital, or the American Embassy or Consulate, to be considered for this service. On the other hand, if death occurs overseas, a Retired person or their family member's body may not be shipped by AMC. Active Duty persons and their families fall under different regulation and will be evacuated to the CONUS.

17. What is "Show Time"? Show Time is a roll call of prospective Space-required and Space-A passengers who are waiting for a specific flight. The total available seats are allocated to travelers based on priority category and date/time sign up. See text for details. Failure to make Show Time will result in not making the flight and Show Times can be changed without notice depending on flight arrivals and departures.

18. Why can't passengers arriving at the terminal after Show Time for a flight be processed for that flight? Passengers should realize that many tasks are performed before a flight departs. Every possible effort will be made to process passengers arriving after Show Time if it doesn't jeopardize the aircraft's departure time or mission safety.

19. Are there special eligibility requirements for pregnant women and infants? Yes. Children must be older than 6 weeks to fly on military aircraft. If the infant is younger than 6 weeks old, there must be written permission from a physician to fly for mother and child. Pregnant women may fly without approval until their 34th week of pregnancy. In a medical emergency, a pregnant woman of more than 34 weeks, or a child younger than 6 weeks and the mother will be flown on a medical evacuation (MEDEVAC) flight as patients.

20. What is the scope of the DoD student travel program? Dependent students who attend school in the United States are authorized one round-trip travel per fiscal year from the school location to the parents' duty station overseas, including Alaska and Hawaii. The student travel program began in 1984 as a quality of life initiative for service members stationed overseas who had children attending secondary or undergraduate school in the United States. The plan has fluctuated over the years. The rule for the travel program apply to service members permanently assigned outside CONUS authorized to have family members reside with them. The student dependent must be unmarried, under 23 and pursuing a secondary or undergraduate education and possess a valid DD Form 1173 ID card.

21. What is the Environmental and Morale Leave (EML) Program? This program is designed to provide environmental relief from a duty station which has some "drawbacks" and to offer a source of affordable recreation otherwise not available. In simple terms, it boils down to allowing Active Duty military personnel and their dependents to fly Space-A on military aircraft. There are, however, a couple of big differences in EML leave and regular Space-A leave. **First, dependents are permitted to travel accompanied or UNACCOMPANIED by their sponsor.** They may utilize "suitably equipped DoD logistic-type aircraft" as well as AMC channel and contract aircraft. **Secondly, EML has a Category 2 classification which is higher than regular Retired Space-A classification (Category 4).** EML is in the same general travel category afforded Active Duty personnel on leave. Military sponsors and/or dependents on EML revert to ordinary leave status when they arrive in CONUS. They regain their EML status only when they depart CONUS for their EML program area. A good bit of EML travel is utilized in the Middle and Far East areas. This means that fewer flights may be available from this area for lower category personnel. The EML program is a tremendous morale booster to those assigned in far-off places and is very popular in these areas.

"Oh yeah? My dad's got the revalidation ribbon with six oak leaf clusters."

22. My husband was killed in Vietnam and is buried in the Punch Bowl (National Memorial Cemetary of the Pacific) in Hawaii. The children and I would like to take a trip to Hawaii to visit his grave. Can we fly in a Space-A status? No. Sorry, but widow/ers are not afforded the privilege of Space-A air travel. The rules state that family members must be accompanied by their military sponsor, so naturally this is impossible. **There have been proposals advanced, namely by the National Association of Uniformed Services/Society of Military Widows (NAUS/SMW) and others, to support a change to the DoDD which prohibits widow/ers of uniformed personnel from using overseas (and any other) Space-A travel.**

23. May I register (sign-up) by FAX, letter/courrier, or in person at the same departure terminal more than one time for five different foreign countries in order to improve my chances for selection to a particular country? Space-A passengers may have only one registration (sign-up) record at a passenger terminal, specifying a maximum of five countries (the fifth country may be "All" in order to allow the widest opportunity for Space-A air travel). This record may be changed at any time to include adding or deleting countries to which a passenger want to travel, but the Julian date and time will be adjusted to the date of the change. No passenger may have two or more records with separate information. However, you may sign-up at several departure terminals in order to improve your chances for selection for air travel. For example, in the Washington, DC/Philadelphia area you can sign-up at McGuire AFB, Philadelphia IAP, Andrews AFB, and Dover AFB for Air Travel to Central Europe.

"Hello. Twenty-fifth Battalion? This is Corporal Atkins in CONUS. I'll have to extend my leave. All overseas flights have been cancelled."

24. What happens to your sign-up records at a departure location when you fly from that station? Note carefully that once passengers are selected for a flight, their name **will be removed from the standby register for all destinations.**

25. May a Retired service member, who relies on a guide dog because of vision deficiency, travel with the animal aboard military aircraft Space-A? Yes. This is allowed when the dog is properly harnessed and muzzled and the animal does not obstruct the aisle. Also, the dog may not occupy a seat in the aircraft.

26. May pets be transported Space-A? Not by Space-A passengers. Active Duty personnel may move pets Space-A on military contract flights when the sponsor is traveling on a permanent change of station.

27. What documents are required for traveling Space-A? All travelers require a uniformed services ID card. Dependent family members and Retirees require a passport, in most cases. Visas may be required for passport holders. In some cases immunization records are required. See Section IV for detailed requirements.

28. I am retired. When I was on Active Duty, my personnel officer issued me travel and leave orders which specified travel documents and other requirements for visiting foreign countries. Where can I now get that information? Appendix B - Personnel Entrance Requirements in **Military Living's Military Space-A Air Opportunities Around The World** book. You may also check the latest changes to the DoD Foreign Clearance Guides at local personnel offices, AMC Space-A counter or other air departure locations.

29. As a Space-A passenger, will I be subjected to security screening prior to boarding a flight? Yes. In most cases you and your baggage will receive electronic and/or personal security screening prior to boarding the flight or entering a secure area for aircraft boarding.

30. May adult family members who are dependent children because of a handicap or are permanently disabled, and who have a valid DD Form 1173 military ID card, travel with their sponsor regardless of age? Yes. They may travel on the same basis as any other dependent on flights going overseas and in the overseas theater.

FEES

31. What fees will Space-A passengers be required to pay? All passengers departing CONUS, Alaska, or Hawaii on a commercial aircraft, from a commercial airport, must pay a $6 departure tax that goes toward airport improvements. Also, all Space-A passengers departing on commercial contract mission **to the United States** must pay a $5 immigration inspection fee, a $5 customs inspection fee, and a $2 agriculture inspection fee. Some foreign departure terminals may also collect a departure tax, e.g., $25 AU when leaving Australia.

FOOD AND BEVERAGE SERVICE

32. Is food served to Space-A passengers on the flight? Food and soft drinks are FREE on AMC contract flights. There is a charge if Space-A passengers want to eat on other flights. You can purchase healthy heart menus from the in-flight kitchen. The snack menu, at $1-1.50, includes sandwich, salad or vegetables, fruit and milk or soft drink. The breakfast menu, at $1-1.50, includes cereal or bagel, fruit, danish, and milk or juice. The sandwich meal, at $2.50-3, includes sandwich, fruit, vegetable or salad, snack or dessert, milk, juice or soft drink. These meals are served at the appropriate time in the flight. Reservations are made at the time of seat assignment or other times in the flight processing. You may bring your own snacks (food) aboard. **New prices are established on 1 October each year.**

"Sorry sir. You should have revalidated this free refill receipt by seventeen hundred hours."

33. Are specialized meals available to Space-A passengers? Specialized meals are made available for duty passengers only for medical or religious reasons. If you need special food, we suggest you bring your own to maintain flexibility. Check with the Air Passenger Terminals regarding any restrictions on carrying food aboard, as this can differ from place to place. While you can make your requirements known to passenger processing personnel at the time of flight processing, the chanceof having additional specialized meals available at the last minute for passengers might be slim.

34. How are alcoholic beverages handled? Alcoholic beverages are not served on military aircraft. All open (seals broken) containers of alcoholic beverages will be confiscated if on your person or in your carry- on baggage. In many cases, sealed alcoholic containers may be checked. Check with the Air Passenger Terminal for more information. You may not consume alcoholic beverages from your own supply on a military aircraft. AMC commercial contract flights, which frequently carry Space-A passengers, offer beer and wine to everyone of legal age for a fee.

"Looks like everything's changed except SOS, Willie."

35. How is food service handled on USN, USMC, USCG, USAF (USAFR, USAG) and other non-AMC flights? Most departure terminals have food service for crews and passengers. If the flight duration is more than approximately four hours, you will be notified in time to obtain your own box of food and drinks. Most flights have coffee and tea and all flights have drinking water on board.

CHANCES OF FLYING SPACE-A

36. How about Space-A availability? Space-A air opportunities change daily and, in fact, even hourly. There are over 325 very active locations at which uniformed personnel their eligible family members and others may fly Space-A to other locations. There are also many other less active locations which offer some Space-A air opportunities. We estimate that more than 800,000 Space-A flights (all services) are taken every year. Availability is subject to: time of the year, needs of the military services, quantity of flights, frequency of flights, and the number of people attempting to fly Space-A.

37. What is the best time of the year to travel Space-A? The best time is a function of: departure locations, arrival locations, space-required needs and the number of people waiting for Space-A transportation. Generally the best times to travel Space-A are autumn, late winter, early spring, and after 15 July. It is best to avoid travel between 1-5 January, 1 May-15 July, 15-30 November and 15-25 December when traffic is heaviest.

38. Who flies Space-A the most - enlisted personnel, officers, retired members or dependents? Enlisted members travel Space-A more than the others.

39. Which uniformed service used Space-A more than the others? Air Force members travel Space-A more than members from any other service, followed by the U.S. Army, U.S. Navy and U.S. Marine Corps.

PRIORITY FOR SPACE -A TRAVEL

40. Who has priority on Space-A flights? The DoD has established a priority system for allocating Space-A air travel. **This system is described in detail in Chapter 4, Space-A Passenger Regulations, DoDD 4515.13-R (Draft).** The general categories and their travel priorities are as follows:

Category 1: Emergency travel and Hostile Fire, Environmental and Morale Leave (EML) travel.

Category 2: EML, DoD Dependent School (DoDDS) teachers on EML, uniformed services personnel, foreign exchange military personnel, close blood or affinitive relatives of service members, civilian employees stationed overseas, dependents of service members and civilians, including college students up to age 23, cadets and midshipmen of the U.S. Military and Coast Guard Academies, including foreign military cadets and midshipmen attending these academies, U.S. civilian patients, Medal of Honor recipients and accompanying dependents, military personnel traveling on non-chargeable leave, e.g., house hunting trips, un-accompanied dependents (17 years plus) on EML and DoDDS teachers or their dependents on EML summer break.

Category 3: Secondary school students, dependent children of members of the uniformed services, DoD and other civilian employees, trade or college students of members serving overseas, dependents joining sponsors overseas, military members traveling on permissive/no-cost Temporary Duty Orders (TDY), dependents acquired overseas, dependents overseas traveling to CONUS to take military academy testing.

Category 4: Retired members of uniformed services and accompanying dependents, Active Duty and "Grey Area" Reservists traveling in the US and its possessions, senior level ROTC. For complete listing of **Space-A Eligible Passengers and Required Documentation, see Section IV of this book.**

"I'd like to toast the military air transport that started us on this trip."

41. May any eligible passenger make reservations for Space-A travel?
No. Space-A passengers may not make reservations and are not guaranteed seats.

The registration for Space-A travel is not a reservation. The DoD is not obligated to continue Space-A passengers' travel or to return them to their point of origin.

42. Does rank/grade have anything to do with who gets a Space-A flight? No. Travel opportunities are available on a first-in first-out basis within DoD established categories and travel priorities. Travel is afforded on an equitable basis to officers, enlisted personnel, DoD/other civilian employees and their dependents without regard to rank or grade, military or civilian, or branch of service.

" Can you land a one forty-one on a carrier?"

43. Are there any circumstances under which a Retired service member in Category 4 may be upgraded to a higher category? You bet there are. If you are traveling Space-A overseas and an emergency occurs at home, you may be upgraded to Category 1, Emergency Travel by the installation commander or his representative under Chapter 4, paragraph 4-3, DoDD 4515.13-R. However, you should have the emergency verified, in writing, by the Red Cross before attempting to obtain an upgrade.

TEMPORARY DUTY AND SPACE-A TRAVEL

44. May uniformed services personnel on official temporary duty orders (TDY) elect to travel Space-A to the TDY point (station)? No. Uniformed services personnel on official TDY orders must travel in a duty status to the TDY point and return.

"The regulation states that you must be requested to come to Guantanamo by someone you know there."

45. Is there any way family members can travel Space-A to their sponsor's TDY point? No. Family members are not authorized Space-A to and from a sponsor's TDY point. TDY personnel may not travel Space-A between their duty station and TDY point as a means to have their dependents travel with them.

46. Can the service member take leave and travel Space-A from the TDY point? Upon arrival at the TDY point, personnel must conduct their business in a TDY status. They may then take ordinary leave while at the TDY point and travel Space-A from the TDY point to another location, but leave must be terminated prior to return travel from the TDY point of origin to the service member's duty station or next TDY location.

47. May family members travel Space-A when the sponsor takes leave at the TDY point? Family members may join the sponsor at the TDY point (at their own expense) in order to travel Space-A with the sponsor while they are on leave.

48. May the service member and dependents travel Space-A between CONUS and overseas? When the service member's permanent duty station and TDY location are within CONUS, Space-A travel to an overseas area and return is authorized. Also, when the service member's duty station and TDY location are overseas, Space-A travel to CONUS and return is authorized. (Note: Dependents may not travel Space-A within CONUS.)

49. When the service member's duty station and TDY location are in different countries overseas, and the service member travels Space-A to CONUS, may they return Space-A to their duty station? No. The service member must return Space-A from CONUS to the overseas TDY point or to a location other than the permanent duty station. He must return to the TDY point (at personal expense, if necessary, if Space-A travel is not possible to the TDY point) in order to complete travel to the permanent duty station in TDY status.

50. What is a simple summary of the above complex guidelines? The bottom line is that service members must always travel between their permanent duty station and a TDY point, or between two TDY points, in a TDY status.

OTHER

51. May Space-A eligible passengers take Space-A air transportation around the world? No. There are insufficient Space-A flights to circumnavigate the earth north to south or south to north. There are adequate flights to travel around the earth east to west or west to east. However, there is one choke point, Diego Garcia, Chagos Archipelago (NKW/FJDG), IO (UK) , through which you are not authorized to travel Space-A. The Secretary of Defense (SECDEF) has limited access to Diego Garcia to mission-essential personnel. Space-A travel through Diego Garcia, including circuitous travel for personnel on official order, is not authorized. This prohibition is found in SECDEF message 250439Z JAN '86 and the DoD Foreign Clearance Guides. Commercial facilities at this UK territory in the Indian Ocean are extremely limited to nonexistent. The Diego Garcia Naval Base does not have lodging, messing and other support facilities essential for non-mission essential travelers.

52. Should I expect to find more than one Space-A roster on base? No. Only one Space-A roster shall be maintained on a base, installation, or post. The maintenance of such a roster is the responsibility of the AMC passenger or terminal service activity. If there is no AMC transportation activity, then the base, installation or post commander designates the agency responsible for maintaining the Space-A roster.

Interior passenger section of a C-5A Galaxy. *(Ken Hockman, photographer, U.S. Air Force photo.)*

APPENDIX N: SPACE-A TRAVEL TIPS

Below are some travel tips which we hope you will find useful. **Many were supplied by our R&R Space-A Report® subscribers**

DOCUMENTS

• Carry passports, military IDs, and travelers checks with you, and not in your luggage. Make xerox or photo copies of your ID cards, credit/debit cards, title page of passports, immunization records, title page of international driver's permit, list of travelers' checks, list of baggage contents, and other important documents. Take one copy with you (not in your luggage) and leave one copy at home or office where it is accessible from overseas.

MEDICATIONS

• Take all medications (prescriptions and over-the-counter) in their original (labeled) containers, and take any essential medications with you on the plane/train not in your baggage.

• If possible, and when necessary, it is recommended that you do not take Dramamine until the plane has been in the air for a while.

CLOTHING

• If you are planning to launder clothes, pack a well wrapped (plastic bag) liquid laundry soap as opposed to a powder soap. Woolite works well for hand as well as machine washing, and comes in both liquid and powder form. Note that laundromats overseas may not have a "permanent-press" cycle on the washer/dryer. They are also much smaller than those here.

• Know the climate at your destination. Travel light. In most cases you will be carrying your own bags. Wear wash-and-wear type clothes. Travel in casual clothes that are loose-fitting and comfortable. Plan your wardrobe such that you can take off or add clothes in layers. Always wear a jacket, light-weight or heavy depending on the weather at your destination. Include a light raincoat or all-purpose coat in lieu of the jacket. Always wear comfortable shoes with low heels or no heels. Pack socks, underwear, etc., in plastic bags.

BAGGAGE

• Folding luggage carriers do not count as checked baggage.

• Because of security problems many U.S. bases no longer have lockers available for storing your luggage.

• Consider using soft-sided luggage to get more into each suitcase. Allow space for items you purchase overseas, or take a collapsible suitcase to bring gifts home.

• There are very few porters at European airports and train stations, and there are limited to no porters at Space-A terminals. Pack only what you can carry comfortably. Can you carry your bags for one mile (15-20 minutes) without setting them down? If not, your bags are too heavy. Get a shoulder bag with small outer compartments. The bag must fit under your seat in the aircraft, and it should be stain resistant and waterproof. Never carry one large bag but split travel articles into two bags for ease in carrying. Bring less clothes and more money. As a general rule, pack a first time, and then cut your original amount of clothes in half and repack.

• Always put your name and address in the inside of your bags as well as on the outside tag of each bag. If the outside tag is lost, your bag can still be returned to you. Put identifying marks in bright colored tape on the outside of your bags for easy identification. (We have a large "C" on each side of our bags.) Lock your bags to protect against partial loss or to at least slow down the would be thief. Officials have keys for different types of bags to be used in an emergency.

• As said earlier, always lock and strap, if available, every bag. Never pack cash, jewelry, medicine or other valuables or hard to get items in your bags. After you have packed your bags, **never leave them unattended, anywhere, for any reason, at any time period, until they are checked for travel.**

CUSTOMS

• Keep receipts for Value Added Tax refunds and for proof of purchase at U.S. Customs.

AMC PLANES AND FLIGHTS

• In a C-5, your seats are above the cargo area, and the seats are usually nicer. **In a C-141, your seats are in lieu of cargo.** They may be regular seats or red fabric/canvas fold down seats. Avoid seats 1A and 1B in a C-5. They are against the bulkhead and do not recline, as well as being opposite the bathroom(s).

• Boarding may be quite different from commercial airlines. There may be ladders to climb, or passengers might be boarded from the open flight line rather than through an enclosed passenger gate. For these reasons slacks are better than skirts for women.

• Climate in the plane may not be standard. In each type of plane there are hot spots and cold spots. Try to dress in layers for comfort and convenience. The flight crew will supply a small pillow, a blanket, and earplugs (on some flights).

• Planes are usually boarded with DV/VIP's or families first. (May not be followed at all stations/locations.)

• Bring something to read on the plane; something to eat, too. Buy a meal on the flight. The food is good and it gives you something to do during the 8-10 hour flight to Europe.

• Usually there is a DV/VIP lounge in the AMC airport terminal available to O-6 and above and to E-9 of the Armed Services.

MONEY

• Exchange some U.S. currency for the currency of at least your first destination country before you go overseas.

• Distribute travelers' checks among those traveling in your group. Consider travelers' checks in various U.S. denominations ($50 or $100), as well as in foreign currency denominations (French francs, Italian lire, German marks, British pounds sterling, Japanese yen and other Asian currencies).

• If possible, bring foreign coins with you for telephone tips, etc. Bring along U.S. change to use in the vending machines on U.S. bases/installations. Bring a personal check or two to cash at an Officers' Club/NCO Club overseas. (You will need a U.S. club card to cash a check in overseas clubs). When dealing with foreign coinage, watch for non-money coins, e.g., telephone tokens in Italy and UK.

• Bring a pocket calculator to convert local prices into U.S. dollars.

• Border towns will usually accept either country's money.

• Bring along U.S. dollars for the flight home ($12 per person for federal inspection, when departing on a contract mission $3 per person for a small meal). Be prepared to take a commercial flight home, and have enough money for that type of flight.

• Be aware that foreign banks may close early on some days; usually the exchanges at major airports and train stations are open 24 hours a day. Exchange your travelers' checks at banks or exchanges, rather than in stores or restaurants. Hotels and stores tend to charge expensive exchange commissions.

• MasterCard and Visa are widely accepted in Europe, as is American Express. Internationally accepted credit cards can be used for cash advances. Also, carry one or two airline credit cards in case of an emergency. Arrange to have funds sent to you via wire to the bank. For tips and payment for services, carry some foreign currency and coins if available, or **carry new U.S. one dollar bills which are readily accepted by service personnel in foreign countries.**

BILLETING

• Check for hotel/motel accommodations at the tourist offices at main train stations and airports. There are also computer matching services at these locations that will provide a list of accommodations, base or location, price range, length of stay and your needed accommodation.

• Your room rate will most often include a continental breakfast,

• The room rate will vary according to:
Class of hotel (Deluxe, First Class, Second Class, etc.);
Type of accommodations (Double bed, King-size bed);
With or without toilet (W/C) in room;
With or without bath/shower in room;
Whether or not the hotel has a restaurant;
Whether or not the hotel has a parking lot;
Whether or not the hotel has an elevator (lift).

• In Great Britain, area libraries usually have a list of local Bed and Breakfast ("B&B") establishments.

• Consider traveling before or after the tourist season in a country, when "in season" rates are no longer in effect. Watch out for trade or other seasonal fares/events that will tie up a large number of hotel rooms.

• Address and telephone numbers (800) can be obtained from the research section of your local library.

• Write foreign country's Tourist Office in the U.S, most are located in New York City and other Gateway cities. Addresses and telephone numbers are available in base and public libraries. They will send various kinds of tourist information as well as hotel/motel price lists.

• Look for different types of accommodations such as "Bed and Breakfasts" or "Pensiones".

TRANSPORTATION

• European and most Asian transportation runs "on time"!

• Use the local public transportation system whenever and wherever possible. Note that there is usually a "smoking" and "non-smoking" section on public transportation.

• Public transportation (buses/subways) usually accepts exact change only. You may have to buy a ticket before boarding, but frequently no one collects bus or tram tickets from you (i.e., the Frankfurt, GE light rail system). **Note:** Do not fail to buy and retain your ticket. The fine for not buying a ticket is extremely high.

• Look for special tourist rates or tourist passes offered by your hotel or the local tourist office. Ask about special transportation rates for round trip travel or time-limited travel, i.e., weekend, five-, seven-, or fifteen-day passes.

• Most European train stations and airports are open 24 hours a day.

• Note the difference between First Class and Second Class on trains. Trains in Europe are heavily utilized, and Second Class may be jammed with students and vacationers during holiday times (Easter, Christmas, New Year's, school breaks, etc.).

• Note that in most cases you can reserve a seat on a train, etc., especially if you want a widow seat or a seat facing in the same direction in which the train is traveling (many seats are reversible).

• As in the U.S., food and drink aboard a train or boat is expensive. You may want to bring your own snack, drink, or lunch on board.

• Be aware of the different fare structures, e.g., a special rate for children, **military,** and animals.

• Points on rental cars: Check AAFES rentals. Rent away from the airport to save money and use low rental agencies, return car full of gas, your insurance may cover the rental car, consider driveaways, you pay a refundable deposit, they put in first tank of gas and you put in the rest. Rental car reservations are essential in most foreign countries. You can make reservations from the USA for major car rental companies. Check any car damage carefully.

LOCAL CUSTOMS

• Know if a visa is necessary for your entry into or exit from the country. See our Military Living's **Military Space-A Air Opportunities Around the World** book or the DOD Foreign Clearance Guide(s) at AMC Space-A counters or military personnel offices which issue worldwide travel orders.

• Know what language is spoken in the part of the country you are visiting, e.g., Switzerland has no official language of its own; rather, the Swiss speak a Swiss-German in the north, French near Geneva, Italian in the south (and English everywhere).

• Bring an English (Foreign Language) dictionary with you. Try to learn the "basics" in the appropriate foreign language, e.g., "Hello," "Goodbye," "Please," "Thank You," "Good Morning," "Good Evening," "Yes," "No," "One," "Two," "Toilet," "train station', "restaurant" etc.

• Study the local customs and manners in the country you plan to visit. For example, know when to shake hands, how to greet a guest, and when to ask for the menu.

• Restaurant menus are often available in English; ask the waiter or hostess for an English-language menu.

• Know when the local and national holidays are in the country you are visiting. Know the stores that are open late. **Get a local map and mark the location of your hotel on it, and have memorized or written down the address where you are staying.**

• Look for an English-speaking tour. You'll get more out of it if the guide does not have to translate into multiple languages.

• Plan to visit the countryside, not just the big cities.

• Note the time differences between where you are and the East Coast of the U.S. especially true in late April and October when our time changes. Typically, there are a telephone, telegraph, and post office located in one central and several other locations.

• Be sure to send postcards and other mail to the U.S. via Air Mail.

OTHER TIPS

• Travel Preparation Time Schedule - Ninety (90) days before departure: Documents, health, language training, guide books and maps, money requirements, travelers' checks. Sixty (60) days before departure: Documents: ID, passport, visas, international driver's permit, immunizations. Thirty (30) days before departure: Health insurance, money. Seven (7) days before departure: Clothes, insurance, luggage, medicines, glasses, film, audio/video tape.

• The successful Space-A traveler has time, patience, funds and is flexible in all aspects of travel.

• Have a map of your destination area for orientation purposes and to avoid becoming lost. It is also useful for measuring local travel distances and paying fares.

• Be flexible in selecting your Space-A route. A direct line to your desired destination may not be the only route to your destination. If possible, select a place with frequency scheduled departures to your planned destination.

• When leaving your car at a departure location, be aware that you may not be able to return via Space-A to your car's location.

• Some bases are more fun than others; try to pick a fun and inexpensive base if you expect to wait for a few days before obtaining a flight.

• Get information from libraries, book and map stores, tourist offices (state, regional and national), travel agents, uniformed services personnel and their families and friends about your destination.

- NOTES -

APPENDIX O: SPECIAL TRAVEL AIDS

Below are some special travel aid publications which we have found to be the most useful to Space-A air travelers.

1. TIPS: CAR RENTAL, HOTEL, OVERSEAS TRAVEL, PACKING, HOLIDAY TRAVEL, AND AIR TRAVEL FOR PETS (include a self-addressed stamped #10 envelope with 75 cents postage), American Society of Travel Agents, Public Relations Department, 1101 King Street, Alexandria, VA 22312. Comm: 703-739-2782. **Free.**

2. FLY RIGHTS, A GUIDE TO AIR TRAVEL IN THE U.S., Department of Transportation, Consumer Information Center, Dept 165-P, Pueblo, CO 81009, **$1.**

3. INTERNATIONAL ASSOCIATION FOR MEDICAL ASSISTANCE TO TRAVELERS (IAMAT), 417 Center Street, Lewiston, NY 14092-3633. Comm: 716-754-4883. This is a no-membership-fee organization that publishes a directory of English-speaking physicians at IAMAT Centers in hundreds of cities all over the world (including some very remote places). For a tax deductible (we believe) contribution of your choice they will provide you with the directory, a passport-sized medical record to be completed by your physician before departure, world immunization charts, and world climate charts. The centers will have daily lists of approved doctors available on a 24-hour basis. The schedule of fees (as of July 1994) agreed to be charged is : office call $45; house/hotel call $55; night, Sunday and local holiday call $65. These fees do not apply to consultations, hospital or laboratory fees. **Free (Donation requested).**

4. THE INTERNATIONAL DIRECTORY OF ACCESS GUIDES, Rehabilitation International USA Inc., 20 West 40th Street, New York, NY 10018. **Free.**

5. NEW HORIZONS FOR THE AIR TRAVELER WITH A DISABILITY, Department of Transportation, Office of Consumer Affairs, 400 Seventh St. S.W., Washington, D.C. 20510. Tel: 202-366-2220. **Free.**

6. DIRECTORY OF USO's WORLDWIDE, United Service Organization (USO) World Headquarters, 601 Indiana Ave NW, Washington, DC 20004, Comm: 202-783-8117. **Free.**

7. FOREIGN ENTRY (VISA) REQUIREMENTS, pub-M-264; **A SAFE TRIP ABROAD,** pub-9493; **TRAVEL TIPS FOR OLDER AMERICANS,** pub-9309; **TIPS FOR AMERICANS RESIDING ABROAD,** pub-9745; **TRAVEL**

WARNING ON DRUGS ABROAD, pub-9558, also publications providing tips for travel to many specific destinations including the Caribbean, Cuba, Central & South Amercia, South Asia, the Middle East & North Africa, Saudi Arabia, Russia, The Peoples Republic of China and Mexico. Department of State, Bureau of Consular Affairs, available at any passport office **free** or order from Superintendent of Documents, GPO, Washington, DC 20402 ($1 each).

8. HEALTH INFORMATION FOR INTERNATIONAL TRAVELERS 1993, HHS pub. no. (CDC) 89-8280, U.S. Department of Health and Human Services, Public Health Service, Center for Disease Control, order from Superintendent of Documents, GPO, Washington, DC 20402, Comm: 202-783-3238, **$6.00.**

9. KNOW BEFORE YOU GO, CUSTOMS HINTS FOR RETURNING RESIDENTS, pub-512; **IMPORTING A CAR,** pub-520; **U.S. CUSTOMS HIGHLIGHTS FOR GOVERNMENT PERSONNEL,** pub-518, **PETS, WILDLIFE** pub-259-187 Department of the Treasury, U.S. Customs Service, available at any passport office **free** or order from Superintendent of Documents, GPO, Washington, DC 20402 ($1 each).

10. CAR RENTAL GUIDE, Facts for consumers Federal Trade Commision, Washington, DC 20580. Tel: 202-326-3650. **Free.**

11. WASHINGTON POST TRAVEL INFO, by FAX (24 hours a day) 1-800-945-5190 with major credit card. See the Sunday Travel Section for a list and code numbers of available articles.

12. INTERNATIONAL DRIVE PERMIT (IDP), Write or telephone: American Automobile Association (AAA), 8300 Old Courthouse Road, Vienna, VA 22182, Tel: 703-790-2600 for IDP application. Cost of IDP is $10.00. An IDP is required in foreign countires to drive a rental car or any vehicle not registered in your name. The IDP costs $10 and is valid for one year.

13. CONSULAR INFORMATION SHEETS, Information as to the location of the U.S. Embassay or Consulate in the subject country, health conditions, political disturbances, unusual currency and entry regulations, crime and security information, and drug penalties. The above information and "Travel Warnings" may be heard by calling (201) 647-5225 from a touchtone phone. Also available at regional U.S. Passport Agencies, or send #10 self addressed stamped envelope to Citizens Emergency Centerm Bureau of Conusular Affairs, Room 4811, N.S., U.S. Department of State, Washington, DC 20520-4818. **Free.**

APPENDIX P: PASSPORTS

The required personnel entry documents vary from country to country, but you probably will require one or more of the following: passport or other proof of citizenship, visa, or tourist card. **A few countries also require evidence that you have enough money for your trip and/or your ongoing transportation tickets.** To find out what you need, consult: (1) This Appendix; (2) Appendix R: Visa Information; (3) Appendix B: Entrance Requirements, found in Military Living's **Military Space-A Air Opportunities Around the World;** and (4) Embassies or nearest consulates of the countries you plan to visit. These official institutions representing foreign governments in the U.S. are located in Washington, DC and major U.S. cities and have the most up-to-date information. They are, therefore, your best source. Consult your local library or Foreign Entry Requirements, Pub M-264 for addresses and telephone numbers.

Who Needs A Passport?: U.S. citizens need passports to depart or enter the U.S. and to enter most foreign countries. Exceptions include short-term travel between the U.S., Mexico and Canada. For many Caribbean countries, a birth certificate or voter registration card is acceptable proof of U.S. citizenship. However, a valid U.S. passport is the best travel documentation available and, with appropriate visas, is acceptable in all countries. With the number of international child custody cases on the rise, several countries have instituted a passport requirement to prevent potential international child abductions. For example, Mexico has a law regarding children traveling alone or with only one parent. If the child travels with one parent, a written notarized consent must be obtained by the other parent. No authorization is needed if the child travels alone and is in possession of a U.S. passport. Children traveling alone with a birth certificate require authorization from both parents.

When To Apply: Demand for passports become heavy in January and begins to decline in August each year. You can help reduce U.S. Government expenses and avoid delays by applying between September and December. However, even in those months, periods of high demand for passports can occur. Apply several months in advance of your planned departure whenever possible. If you need a visa, allow additional time. (Passport agencies will expedite issuance in cases of genuine, documented emergencies.)

How To Apply: For your first passport, you must present, in person, a completed form SDP-11, *Passport Application*, at one of the passport agencies listed in this Appendix or at one of several thousand Federal or State courts or U.S. Post Offices authorized to accept passport applications. Contact the nearest passport agency for the addresses of the passport acceptance facilities in your area. If you have had a previous passport and wish to obtain another, you may be eligible to apply by mail. If you do not qualify, you must apply in person at an authorized office.

What You Need To Obtain A Passport:

(1) A properly completed *Passport Application* (Form DSP-11).

(2) Proof of U.S. Citizenship.

a. Use your previously issued passport or one in which you were included. If you are applying for your first passport or cannot submit a previous passport, you must submit other evidence of citizenship.

b. If you were born in the U.S. you should produce a birth certificate. This must show that the birth record was filed shortly after birth and must be certified with the registrar's signature and raised, impressed, embossed, or multi-colored seal. You can obtain a certified copy from the bureau of Vital Statistics in the state or territory where you were born. (Notifications of Birth Registration or Birth Announcements are not normally accepted for passport purposes.) A delayed birth certificate (one filed more than one year after the date of birth) is acceptable provided it shows a plausible basis for creating this record.

If you cannot obtain a birth certificate submit a notice from a state registrar stating that no birth record exists, accompanied by the best secondary evidence possible. This may include a baptismal certificate, a hospital birth record, affidavits of persons having personal knowledge of the facts of your birth, or other documentary evidence such as early U.S. census, school, family Bible records, newspaper files, and insurance papers. A personal knowledge affidavit should be supported by at least one public record reflecting birth in the U.S.

c. If you were born abroad you can use: 1) A Certificate of Naturalization; 2) A Certificate of Citizenship; 3) A Report of Birth Abroad of a Citizen of the U.S.A. (FS-240); or 4) A Certification of Birth (FS-545 or DS-1350). If you did not have any of these documents and are a U.S. citizen, you should take all available proof of citizenship to the nearest U.S. passport agency and request assistance in proving your citizenship.

(3) Proof of Identity - You must also establish your identity to the satisfaction of the person accepting your application. The following items generally are **acceptable documents of identity** if they contain your signature and if they readily identify you by physical description or photograph: 1) A previous U.S. passport; 2) A certificate of naturalization or citizenship; 3) A valid driver's license; and 4) A government (federal, state, municipal) ID card. **The following documents are not acceptable:** 1) Social Security Card; 2) Learner's or temporary driver's license; 3) Credit card of any type; 4) Any temporary or expired ID card or document; 5) Any document which has been altered or changed in any manner. If you are unable to present one of the first four documents to establish your identity, you must be

accompanied by a person who has know you for at least two years and who is a U.S. citizen or permanent resident alien of the U.S. That person must sign an affidavit in the presence of the same person who executes the passport application. The witness will be required to establish his or her own identity.

(4) Photographs - Present two identical photographs of yourself which must be sufficiently recent (normally taken within the past six months) to be a good likeness. The photographs must not exceed 2" x 2" in size. The image size measure from the bottom of your chin to the top of your head (including hair) must not be less than 1" or more than 1-3/8", with your head taking up most of the photograph. **Passport photographs are acceptable in either black and white or color.** Photographs must be clear, front view, full faced, and printed on thin white paper with a plain white or off-white background. Photographs should be portrait type prints taken in normal street attire without a hat and must include no more than the head and shoulders or upper torso. Dark glasses are not acceptable except when worn for medical reasons. Only applicants who are on Active Duty in the U.S. Uniformed Services and are proceeding abroad in the discharge of their duties may submit photographs in service uniform. (Passport Services encourages photographs in which the applicant is relaxed and smiling.) Newspaper and magazine prints and most vending machine prints are not acceptable for use in passports. **No inclusion of family members in your passport. Since January 1981, all persons have been required to obtain individual passports in their own name. You may not include your spouse or children in your passport.**

When To Apply In Person: You must apply in person for your passport if: (1) This is your first passport; (2) You are under 18 years of age (applicants between the ages of 13 and 18 must appear in person before the clerk or agent executing the application; for children under the age of 13, parents or legal guardians may apply on their behalf), and (3) Your most recent previous passport was issued more than 12 years ago or you were under age 16 when you received it. Applicants who are required to appear in person may do so at a U.S. passport agency or at one of the post offices or clerks of court authorized to accept passport applications. **If you are over 18 and must apply in person, the total charge will be $65 for a 10 year passport; if you are under 18 year of age, you will pay $40 for a 5-year passport. Form DSP-11 is the appropriate form to use.**

When To Apply By Mail: You may apply by mail if: (1) You have been issued a passport with 12 years prior to the date of a new application; (2) You are able to submit your most recent U.S. passport with your new application; and (3) Your previous passport was not issued before your 16th birthday.

How To Apply By Mail: If you are eligible to apply by mail, obtain **Form DSP-82, "*Application for Passport by Mail,*"** from one of the offices accepting applications or from your travel agent, and complete the info requested on the reverse side of the form: (1) Sign and date the application; (2) Attach your previous passport (two identical 2" x 2" photographs which are sufficiently recent, normally taken within the past six month, to be a good likeness, and the $55 passport renewal fee; (3) Mail the completed application and attachments to one of the passport agencies listed in this Appendix. An incomplete or improperly prepared application will delay issuance of your passport.

Payment Of Passport Fees: The following forms of payment are acceptable: (1) Bank draft or cashier's check; (2) Check: certified, personal, travelers (for exact amount); and (3) Money order: U.S. postal, international, currency exchange bank. Cash should not be sent through the mail and is not always accepted by post offices and clerks of court.

Diplomatic & Official Passports: If you are being assigned abroad on U.S. Government business and wish to apply by mail for a no-fee type passport (no-fee regular, official, diplomatic), you must submit the mail-in application form, your authorization to apply for a no-fee passport, your previous passport, and two photographs to the Passport Agency in Washington, DC for processing.

After You Receive Your Passport: Be sure to sign it and fill in the personal notification data page. Your previous passport will be returned to you with your new passport.

Additional Visa Pages: If you require additional visa pages before your passport expires, you can obtain them by submitting your passport to one of the passport agencies listed in **Appendix Q**. You may also request a 48 page passport at the time of application if you are planning to travel abroad frequently.

Validity Of An Altered Or Mutilated Passport: If you mutilate or alter your U.S. passport in any way (other than changing the address and personal notification data) you may render it invalid, cause yourself much inconvenience, and expose yourself to possible prosecution under the law (Section 1543 of Title 22 of the U.S. Code). Mutilated or altered passports should be turned in to passport agents,authorized postal employees, or U.S. consular offices abroad.

Loss Or Theft Of U.S. Passport: Your passport is a valuable document of citizenship and identity which should be carefully safeguarded. Its loss may cause you unnecessary travel complications as well as significant expense. If your passport is lost or stolen in the U.S., report the loss or theft immediately to Passport Services, 1111 19th Street NW, Department of State, Washington, DC 20524-0002, or to the nearest passport agency. Should your passport be lost or stolen abroad, report the loss immediately to the nearest U.S. embassy or consulate and to the local police authorities. If you can provide the consular officer with the information contained in the passport, it will facilitate the issuance of a new passport. Therefore, **we suggest you photocopy the data page of your passport** and keep it in a separate place, or leave the passport number, date, and place of issuance with a relative or friend in the U.S.

Other Passport Information: Sometimes travelers will depart for their intended trip with a passport that is about to expire. Travelers should be aware that there are a number of countries which will not permit visitors to enter and will not place visas in passports which have a remaining validity of less than six months. **If you return to the U.S. with an expired passport, you are subject to a passport waiver fee of $80.** This fee is payable to the Immigration and Naturalization Service at the port of entry. Additional passport information may be obtained from any of the passport agencies listed in this appendix.

Ann and Roy Crawford find their books and maps posted all over the world. This one is at the Keflavik Naval Air Station in Iceland!

APPENDIX Q: PASSPORT AGENCIES

Boston Passport Agency, Room 247, Thomas P. O'Neill Federal Bldg, 10 Causeway Street, Boston, MA 02222. *Recording: (617) 565-6998, **Public Inquiries: (617) 565-6990.

Chicago Passport Agency, Suite 380, Kluczuski Federal Bldg, 230 South Dearborn Street, Chicago, IL 60604-1564. *Recording (312) 353-5426, **Public Inquiries: (312) 353-7155.

Honolulu Passport Agency, Room C-106, New Federal Bldg, 300 Ala Moana Blvd, PO Box 50185, Honolulu, HI 96850. *Recording: (808) 541-1919, **Public Inquiries: (808) 541-1918.

Houston Passport Agency, Suite 1100, Concord Towers, 1919 Smith Street, Houston, TX 77002. *Recording: (713) 653-3159, **Public Inquiries: (713) 229-3600.

Los Angeles Passport Agency, Room 13100, 11000 Wilshire Blvd, Los Angeles, CA 90024-3615. *Recording (213) 209-7070, **Public Inquiries: (213) 209-7075.

Miami Passport Agency, 16th Floor, Federal Office Bldg, 51 Southwest First Avenue, Miami, FL 33130-1680. *Recording: (305) 536-5395 (English) (305) 536-4448 (Spanish), **Public Inquiries: (305) 536-4681/83.

New Orleans Passport Agency, Postal Services Bldg., Room T-12005, 701 Loyola Avenue, New Orleans, LA 70113-1931. *Recording: (504) 589-6728, **Public Inquiries: (504) 589-6161/63.

New York Passport Agency, Room 270, Rockefeller Center, 630 Fifth Avenue, New York, NY 10111-0031. *Recording: (212) 541-7700, **Public Inquiries: (212) 541-7710.

Philadelphia Passport Agency, Room 4426, Federal Bldg, 600 Arch Street, Philadelphia, PA 19106-1684. *Recording: (215) 597-7482, **Public Inquiries: (215) 597-7480/81.

San Francisco Passport Agency, Suite 200, 525 Market Street, San Francisco, CA 94105-2773. *Recording: (415) 974-7972, **Public Inquiries: (415) 974-9941/48.

Seattle Passport Agency, Room 992, Federal Office Bldg, 915 Second Avenue, Seattle, WA 98174-1091. *Recording: (206) 442-7941/43, **Public Inquiries: (206) 442-7945/47.

Stamford Passport Agency, One Landmark Square, Street Level, Stamford, CT 06901-2767. *Recording: (203) 325-4401., **Public Inquiries: (203) 325-3538/39/30.

Washington Passport Agency, 1111 19th St NW, Washington, DC 20524-0002. *Recording: (202) 647-0518, **Public Inquiries: (202) 647-0518.

***The 24-hour recording includes general passport info, passport agency location & hour of operation. **For other questions, call Public Inquiries number.**

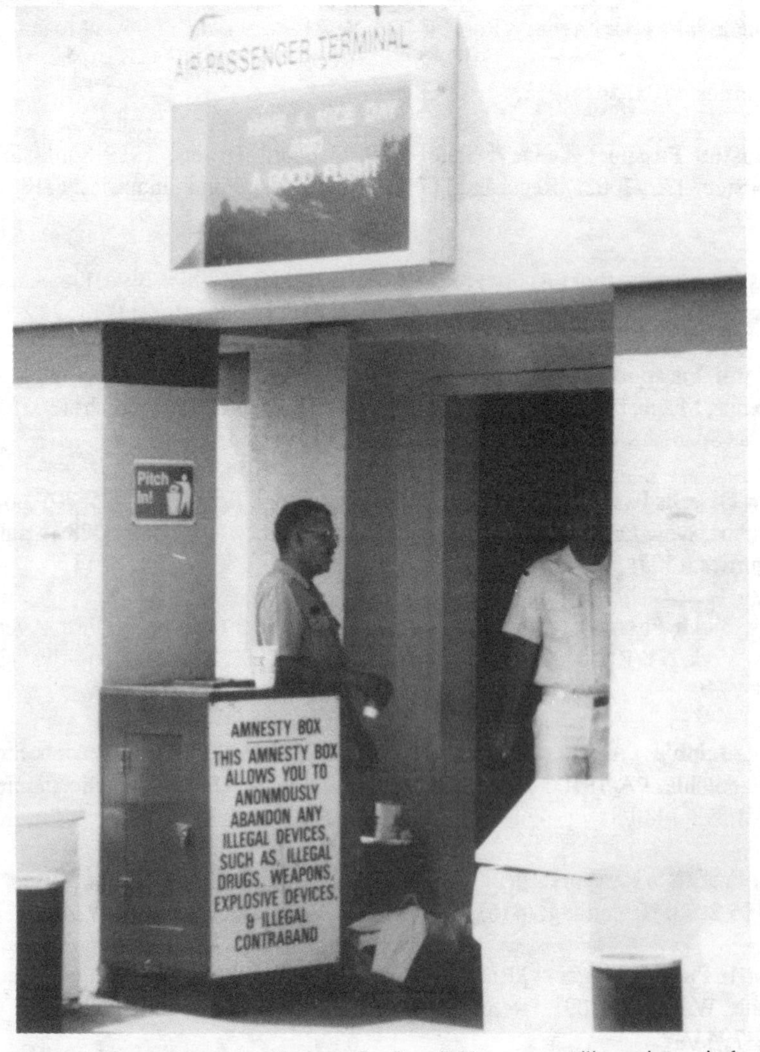

Pay careful attention to the "Amnesty Box" available at most military air terminals.

— Photo by Roy Crawford, Sr.

APPENDIX R: VISA INFORMATION

OBTAINING A FOREIGN VISA

A visa is a permit to enter and leave the country to be visited. It is a stamp of endorsement place in a passport by a consular official of the country to which entry is requested. Many countries require visitors from other nations to have in their possession a valid visa obtained before departing from their home country. A visa may be obtained from foreign embassies or consulates located in the U.S. (Visas are not always obtainable at the airport of entry of the foreign location and verification of visa issuance must be made in advance of departure.) Various types of visas are issued depending upon the nature of the visit and the intended length of stay. **Passport services of the Department of State cannot help you obtain visas.**

A valid passport must be submitted when applying for a visa of any type. Because the visa is usually stamped directly onto one of the blank pages in your passport, you will need to fill out a form and give your passport to an official of each foreign embassy or consulate. The process may take several weeks for each visa, so apply well in advance. The visa requirements of each country will differ.

Some visas require a fee. You may need one or more photographs when submitting your visa applications. They should be full-faced, on white background and should not be larger than 3" x 3" nor smaller than 2.5" x 2.5".

Several countries do not require U.S. citizens to obtain passports and visas for certain types of travel, mostly tourist. Instead, they issue a simple tourist card which can be obtained from the nearest consulate of the country in question (presentation of a birth certificate or similar documentary proof of citizenship may be required). In some countries, the transportation company is authorized to grant tourist cards. A fee is required for some tourist cards.

Some Arab or African countries will not issue Visas or allow entry if your passport indicates travel to Israel or South Africa. Consult the nearest U.S. Passport Agency for guidance if this applies to you.

The official institutions (embassies or consulates) representing foreign governments in the U.S. are located in Washington, DC (see below), and major U.S. cities and have the most up-to-date information. They are, therefore, your best source. Double check visa requirements before you leave. (***The Congressional Directory*, available at most public libraries, lists their addresses and phone numbers.)**

For your convenience, we have listed below the names, addresses and phone numbers of embassies in the countries where stations are frequently used by DoD-owned or controlled aircraft. U.S. Trust Territories & possessions overseas have the same requirements as the U.S.has for U.S. citizens upon return from a foreign country. If you wish to travel to a country not listed below, visa and other personnel entry requirements can be obtained from the "***DoD Foreign Clearance Guides*" available at most AMC (USAF) and other passenger service counters or at many military personnel offices.**

NOTE: Embassies may close on their respective national holidays. Call before going to be sure they are open. We have listed only the Washington, DC based foreign embassies. There may be consulates of these embassies in other major cities which have visa issuing authority.

ANTIGUA-BARBUDA (AN-BD)
Embassy of Antigua and Barduda Intelsat Building, Ste 4M, 3400 International Dr. NW
Washington, DC 20008
Comm: (202) 362-5122/5166/5211

ARGENTINA (AG)
Embassy of the Argentine Republic
1600 New Hampshire Ave NW
Washington, DC 20009
Comm: (202) 939-6400

AUSTRALIA (AU)
Embassy of Australia
1601 Massachusetts Ave NW
Washington, DC 20036
Comm: (202) 797-3000/1-800-242-2878

BAHRAIN (BA)
Embassy of the State Of Bahrain
3502 International Dr NW
Washington, DC 20008
Comm: (202) 342-0741/42.

BARBUDA (BD)
(See Antigua)

BELGIUM (BE)
Embassy of Belgium
3330 Garfield Street NW
Washington, DC 20008
Comm: (202) 333-6900

BELIZE (BZ)
Embassy of Belize
2535 Massachusetts Ave N.W.
Washington, DC 20008
Comm: (202) 332-9636

BERMUDA (BM)
(See United Kingdom.)

BOLIVIA (BO)
Embassy of Bolivia (Consulate Section)
3014 Massachusetts Ave NW
Washington, DC 20008
Comm: (202) 232-4828, or
202-483-4410

BRAZIL (BR)
Brazilian Embassy
(Consular Section)
3009 Whitehaven St. NW
Washington, DC 20008
Comm: (202) 745-2828

CANADA (CN)
Canadian Embassy
501 Pennsylvania Ave NW
Washington, DC 20001
Comm: (202) 682-1740

CHAD (CD)
Embassy of the Republic of Chad
2002 R Street NW
Washington, DC 20009
Comm: (202) 462-4009

CHILE (CH)
Embassy of Chile
1732 Massachusetts Ave NW
Washington DC 20036
Comm: (202) 785-3159

COLUMBIA (CL)
Embassy of Columbia (Consulate)
1825 Connecticut Ave NW
Washington, DC 20009
Comm: (202) 332-7476

COSTA RICA (CS)
Embassy of Costa Rica
(Consulate Section)
1825 Connecticut Ave NW #211
Washington, DC 20009
Comm: (202) 328-6628

CUBA (CU)
Entry to Cuba is permitted only through Guantanamo Bay (U.S. property), and a visa is not required.

CYPRUS (CY)
Embassy of the Republic of Cyprus
2211 R Street NW
Washington, DC 20008
Comm: (202) 462-5772

DENMARK (DN) (including GREENLAND)
Royal Danish Embassy
3200 Whitehaven Street NW
Washington, DC 20008
Comm: (202) 234-4300

DOMINICAN REPUBLIC (DR)
Embassy of the Dominican Republic
1715 22nd Street NW
Washington, DC 20008
Comm: (202) 332-6280

ECUADOR (EC)
Embassy of Ecuador
2535 15th Street NW
Washington DC 20009
Comm: (202) 234-7166

EGYPT (EG)
Embassy of the Arab Republic of Egypt
2310 Decatur Place NW
Washington, DC 20008
Comm: (202) 234-3903

EL SALVADOR (ES)
Embassy of El Salvador
Consulate General of El Salvador
1010 16th Street NW
3rd Floor
Washington, DC 20036
Comm: (202) 331-4032

GERMANY (GE)
Embassy of the Federal Republic of Germany
4645 Reservoir Road NW
Washington, DC 20007
Comm: (202) 298-4000

GREECE (GR)
Embassy of Greece (Consulate Section)
2221 Massachusetts Ave NW
Washington, DC 20008
Comm: (202) 232-8222

GREENLAND (GL)
(See Denmark)

GUATEMALA (GT)
Embassy of Guatemala
2220 R Street NW
Washington, DC 20008
Comm: (202) 745-4952

GUYANA (GY)
Embassy of Guyana
2490 Tracy Place NW
Washington, DC 20008
Comm: (202) 265-6900/03

HAITI (HA)
Embassy of Haiti
2311 Massachusetts Ave NW
Washington, DC 20008
Comm: (202) 332-4090

HONDURAS (HO)
Embassy of Honduras (Consular Section)
1511 K Street NW, Suite 927
Washington, DC 20005
Comm: (202) 223-0185

HONG KONG
(See United Kingdom)

ICELAND (IC)
Embassy of Iceland
2022 Connecticut Ave NW
Washington DC, 20008
Comm: (202) 265-6653/55

INDONESIA (IE)
Embassy of the Republic of Indonesia
2020 Massachusetts Ave NW
Washington, DC 20036
Comm: (202) 775-5200

IRELAND (IR)
Embassy or Ireland
2234 Massachusetts Ave NW
Washington, DC 20008
Comm: (202) 462-3939

ISRAEL (IS)
Embassy of Israel
3514 International Dr NW
Washington, DC 20008
Comm: (202) 364-5500

ITALY (IT)
Embassy of Italy (Consulate)
1601 Fuller St., NW
Washington, DC 20009
Comm: (202) 328-5500

JAMAICA (JM)
Embassy of Jamaica
1850 K Street NW #355
Washington, DC 20006
Comm: (202) 452-0660

JAPAN (JA)
Embassy of Japan
2520 Massachusetts Ave NW
Washington, DC 20008
Comm: (202) 939-6800

JORDAN (JR)
Embassy of the Hashemite Kingdom of Jordan
3504 International Dr NW
Washington, DC 20008
Comm: (202) 966-2664

KENYA (KE)
Embassy of Kenya
2249 R Street NW
Washington, DC 20008
Comm: (202) 387-6101

KOREA (RK)
Embassy of the Republic of Korea
(Consulate Division)
2600 Virginia Ave NW # 208
Washington, DC 20037
Comm: (202) 939-5660/63

LIBERIA (LI)
Embassy of the Republic of Liberia
5201 16th Street NW
Washington, DC 20011
Comm: (202) 723-0437/40

MALAYSIA (MA)
Embassy of Malaysia
2401 Massachusetts Ave NW
Washington, DC 20008
Comm: (202) 328-2700

MARSHALL ISLANDS, REPUBLIC OF (MI)
Representative Office
1901 Pennsylvania Ave NW #1004
Washington, DC 20006
Comm: (202) 234-5414

MEXICO (MX)
Embassy of Mexico (Consular Section)
2827 16th Street NW
Washington, DC 20009-4260
Comm: (202) 736-1000

MICRONESIA, FEDERATED STATES OF (FM)
Embassy of the Federated States of Micronesia
1725 N Street NW

Washington, DC 20036
Comm: (202) 223-4383

NETHERLANDS (NT)
Embassy of the Netherlands
4200 Linnean Ave NW
Washington, DC 20008
Comm: (202) 244-5300

NEW ZEALAND (NZ)
Embassy of New Zealand
37 Observatory Circle NW
Washington, DC 20009
Comm: (202) 328-4800

NICARAGUA (NI)
Embassy of Nicaragua
1627 New Hampshire Ave NW
Washington, DC 20008
Comm: (202) 939-6531/34

NIGER (NG)
Embassy of the Republic of Niger
2204 R Street NW
Washington, DC 20037
Comm: (202) 483-4224

NORWAY (NO)
Royal Norwegian Embassy
2720 34th Street NW
Washington, DC 20008
Comm: (202) 333-6000

OMAN (OM)
Embassy of the Sultanate of Oman
2342 Massachusetts Ave NW
Washington, DC 20008
Comm: (202) 387-1980/82

PALAU, REPUBLIC OF (PL)
Representative Office
444 North Capitol Street #308
Washington, DC 20008
Comm: (202) 624-7793

PANAMA (PN)
Embassy of Panama
2862 McGill Terrace NW
Washington, DC 20008
Comm: (202) 483-1407

PARAGUAY (PG)
Embassy of Paraguay
2400 Massachusetts Ave NW
Washington, DC 20008
Comm: (202) 483-6960

PERU (PE)
Embassy of Peru
1700 Massachusetts Ave NW
Washington, DC 20036
Comm: (202) 833-9860/69

PHILIPPINES (RP)
Embassy of the Philippines
1617 Massachusetts Ave NW
Washington, DC 20036
Comm: (202) 483-1533

PORTUGAL (PO)
(includes the Azores)
Embassy of Portugal
2310 Tracey Place NW
Washington ,DC 20008
Comm: (202) 322-3007

SAUDI ARABIA (SA)
The Royal Embassy of Saudi Arabia
601 New Hampshire Ave NW
Washington, DC 20037
Comm: (202) 333-4595

SENEGAL (SN)
Embassy of the Republic of Senegal
2112 Wyoming Ave NW
Washington, DC 20008
Comm: (202) 234-0540

SINGAPORE (SG)
Embassy of Singapore
1842 R Steet NW
Washington, DC 20009
Comm: (202) 667-7555

SOLOMON ISLANDS (SI)
Permanent Mission of the Solomon Islands to the U.S.
820 Second Ave, Ste 800A
New York, NY 10017
(202) 462-1340

SOMALIA (SM)

Consulate of the Somalia Democratic Republic
New York, NY
Comm: (212) 688-9410

SPAIN (SP)

Embassy of Spain
2700 15th Street NW
Washington, DC 20009
Comm: (202) 265-0190/91

SUDAN (SU)

Embassy of the Republic of the Sudan
2210 Massachusetts Ave NW
Washington, DC 20008
Comm: (202) 338-8565/70

SURINAME (SR)

Embassy of the Republic of Suriname
4301 Connecticut Ave NW
#108
Washington, DC 20008
Comm: (202) 244-7488

THAILAND (TH)

Embassy of Thailand
2300 Kalorama Road NW
Washington, DC 20008
Comm: (202) 234-5052

TURKEY, (TU)

Enbassy of the Republic of Turkey
1714 Massachusetts Ave NW
Washington, DC 20036
Comm: (202) 659-0742

UNITED KINGDOM (UK)

British Embassy (Consulate Section)
19 Observatory Circle NW
Washington, DC 20008
Comm: (202) 896-0205

URUGUAY (UG)

Embassy of Uruguay
1918 F Street NW
Washington, DC 20008
Comm: (202) 331-1313/16

VENEZUELA (VE)

Embassy of Venezula (Consulate)
1099 30th Street NW
Washington, DC 20007
Comm: (202) 342-2214

ZAIRE (ZA)

Embassy of the Republic of Zaire
1800 New Hampshire Ave NW
Washington, DC 20009
Comm: (202) 234-7690/91

APPENDIX S: CUSTOMS AND DUTY

DECLARATIONS: You **must declare** all articles acquired abroad and in your possession at the time of your return. This includes: **1**) Articles that you purchased; 2) Gifts presented to you while abroad, such as wedding or birthday presents; **3**) Articles purchased in the duty-free shops; **4**) Repairs or alterations made to any articles taken abroad and returned, whether or not repairs or alterations were free of charge; **5**) Items you have been requested to bring home for another person; and **6**) Any articles you intend to sell or use in your business. In addition, you must declare any articles acquired in the U.S. Virgin Islands, American Samoa, or Guam and not accompanying you at the time of your return. The price actually paid for each article must be stated on your declaration in U.S. currency or its equivalent in the country of acquisition. If the article was not purchased, obtain an estimate of its fair retail value in the country in which it was acquired. **Note: The wearing or use of any article acquired abroad does not exempt it from duty. It must be declared at the price you paid for it.** The customs officer will make an appropriate reduction in its value for significant wear and use.

Oral Declarations: Customs declarations forms are distributed on vessels and planes and should be prepared in advance of arrival for presentation to the immigration and customs inspectors. **Fill out the ID portion of the declaration form.** You may declare orally to the customs inspector the articles you acquired abroad if the articles are accompanying you, and you have not exceeded the duty-free exemption allowed. A customs officer may, however, ask you to prepare a written list if it is necessary.

Written Declaration: A written declaration will be necessary **when: 1**) The total fair retail value of articles acquired abroad exceeds you personal exemption; **2**) More than 1 liter (33.8 fl oz) of alcoholic beverages, 200 cigarettes (one carton), or 100 cigars are included; **3**) Some of the items are not intended for your personal or household use, such as commercial samples, items for sale or use in your business, or articles you are bringing home for another person; **4**) Articles acquired in the U.S. Virgin Islands, American Samoa, or Guam are being sent to the U.S.; **5**) A customs duty or internal revenue tax is collectible on any article in your possession; **6**) A customs officer requests a written list; and **7**) If you have used your exemption in the last 30 days.

Family Declaration: The head of a family may make a joint declaration for all members residing in the same household and returning together to the U.S. Family members making a joint declaration may combine their personal exemptions, even if the articles acquired by one member of the family exceed the personal exemption allowed. **Infants and children returning to the U.S. are entitled to the**

same exemptions as adults (except for alcoholic beverages). Children born abroad, who have never resided in the U.S., are entitled to the customs exemptions granted nonresidents.

WARNING! If you understate the value of an article you declare, or if you otherwise misrepresent an article in your declaration, you may have to pay a penalty in addition to payment of duty. Under certain circumstances, the article could be seized and forfeited if the penalty is not paid. It is well know that some merchants abroad offer travelers invoices or bills of sale showing false or understated values. This practice not only delays your customs examination, but can prove very costly. **If you fail to declare an article acquired abroad, not only is the article subject to seizure and forfeiture, but you will be liable for a personal penalty in an amount equal to the value of the article in the U.S. In addition, you may also be liable to criminal prosecution.** Don't rely on advice given by persons outside the Customs Service. It may be bad advice which could lead you to violate the customs laws and incur costly penalties. If in doubt about whether an article should be declared, always declare it first and then direct your question to the customs inspector. **If in doubt about the value of an article, declare the article at the actual price paid (transaction value).** Customs inspectors handle tourist items day after day and become acquainted with the normal foreign values. Moreover, current commercial prices of foreign items are available at all times and on-the-spot comparisons of these values can be made. Play it safe - avoid customs penalties.

YOUR EXEMPTIONS: In clearing U.S. Customs, a traveler is considered either a "returning resident of the U.S." or a "nonresident". Generally speaking, if you leave the U.S. for purposes of traveling, working or studying abroad and return to resume residency in the U.S., you are considered a returning resident by Customs. However, U.S. residents living abroad temporarily are entitled to be classifies as nonresidents, and thus receive more liberal Customs exemptions on short visits to the U.S., provided they export any foreign-acquired items at the completion of their visit. Residents of American Samoa, Guam or the U.S. Virgin Islands, who are American citizens, are also considered as returning U.S. residents. Articles acquired abroad and brought into the U.S. are subject to applicable duty and internal revenue tax, but as a returning resident you are allowed certain exemptions from paying duty on items acquired while abroad.

$400 Exemption: Articles totaling $400 (based on the fair value of each item in the country where acquired) **may be entered free of duty, subject to limitations for liquors, cigarettes & cigars, if: 1)** Articles were acquired as an incident of your trip for personal or household use; **2)** You bring the articles with you at the time of your return to the U.S. and they are properly declared to Customs. Articles purchased and left for alterations or other reasons cannot be applied to your $400 exemption when shipped to follow at a later date. The 10% flat rate of duty does not apply to mailed articles. Duty is assessed when received; **3**) You are returning from a stay abroad of at least 48 hours. Example: A resident who leaves U.S. territory at 1:30 PM on June 1st would complete the required 48-hour period at 1:30 PM on June 3rd. This time limitation does not apply if you are returning from Mexico or the U.S. Virgin Islands; **4**) You have not used this $400 exemption, or any part of it, within the preceding 30-day period. Also, your exemption is not cumulative. If you use a portion of your exemption on entering the U.S., then you must wait for 30 days before you are entitled to another exemption other than a $25 exemption; and **5**) Articles are not prohibited or restricted.

Cigars & Cigarettes: Not more than 100 cigars and 200 cigarettes (one carton) may be included in your exemption. Products of Cuban tobacco may be included if purchased in Cuba. This exemption is available to each person regardless of age. Your cigarettes, however, may be subject to a tax imposed by state and local authorities.

Liquor: One liter (**33.8 fl oz**) of alcoholic beverages may be included in this exemption **if: 1**) You are 21 years of age or older; **2**) it is for your own use or for use as a gift; and **3**) It is not in violation of the laws of the state in which you arrive. **Note:** Most states restrict the quantity of alcoholic beverages which you may import. Information about state restrictions and taxes should be obtained from the state government, as laws vary from state to state. Alcoholic beverages in excess of the one liter limitation are subject to duty and internal revenue tax. **Shipping of alcoholic beverages by mail is prohibited by U.S. Postal laws.** Alcoholic beverages include wine and beer as well as distilled spirits.

$800 Exemption: If you return directly or indirectly from the U.S. Virgin Islands, American Samoa or Guam, you may receive a customs exemption of $800 (based on the transaction value of the articles in the country where acquired). Not more than $400 of this exemption may be applied to merchandise obtained elsewhere than in these islands. If you are 21 or older, you may bring in free of duty and tax five liters of alcoholic beverages. That's 169 fluid ozs or about 6 & 1/2 fifths. However, at least four liters must be purchased in the islands and at least one liter must have been produced there. Articles acquired in and sent from these islands to the U.S. may be claimed under your duty-free personal exemption if properly declared. Other provisions under the $400 exemption apply.

$25 Exemption: If you cannot claim the $400 or $800 exemption because of the 30-day or 48-hour minimum limitations, you may bring in free of duty and tax articles acquired abroad for your personal or household use if the total fair retail value does not exceed $25. This is an individual exemption and may not be grouped with other members of a family on one customs declaration. You may include any of the following: 50 cigarettes, 10 cigars, 150 ml (4 fl oz) of alcoholic beverages, or

150 ml of alcoholic perfume. Cuban tobacco products brought directly from Cuba may be included. Alcoholic beverages cannot be mailed into the U.S. Customs enforces the liquor laws of the state in which you arrive. Because state laws vary greatly as to the quantity of alcoholic beverages which can be brought in, we suggest you consult the appropriate state authorities. If any article brought with you is subject to duty or tax, or if the total value of all dutiable articles exceeds $25, no article may be exempted from duty or tax.

GIFTS: Bona fide gifts of not more than $50 in fair retail value where shipped can be received by friends and relations in the U.S. free of duty and tax, if the same person does not receive more than $50 in gift shipments in one day. The "day" in reference is the day in which the parcel(s) are received for customs processing. This amount is increased to $100 if shipped from the U.S. Virgin Islands, American Samoa or Guam. These gifts are not declared by you upon your return to the U.S. **Gifts accompanying you are considered to be for your personal use and may be included within your exemption.** This includes gifts given to you by others while abroad and those you intend to give to others after you return. Gifts intended for business, promotional, or other commercial purposes may not be included. Perfume containing alcohol valued at more than $5 retail, tobacco products, and alcoholic beverages are excluded from the gift provision. **Gifts** intended for more than one person may be consolidate **in the same package** provided they are individually wrapped and labeled with the name of the recipient. Be sure the outer wrapping of the package is **marked:** **(1)** Unsolicited gift; **(2)** Nature of the gift; and **(3)** Its fair retail value. In addition, a consolidated gift parcel should be marked as such on the outside with the names of the recipients listed and the value of each gift. This will facilitate customs clearance of your package. If any article imported in the gift parcel is subject to duty and tax, or if the total value of all articles exceeds the bona fide gift allowance, no article may be exempt from duty or tax. If a parcel is subject to duty, the U.S. Postal Service will collect the duty plus a handling charge in the form of "Postage Due" stamps. **Duty cannot be prepaid. You as a traveler, cannot send a "gift" parcel to yourself nor can persons traveling together send "gifts" to each other.** Gifts ordered by mail from the U.S. do not qualify under this duty-free gift provision and are subject to duty.

OTHER ARTICLES (FREE OF DUTY OR DUTIABLE): Duty preferences are granted to certain developing countries under the Generalized System of Preferences (GSP). Some products from these countries have been exempted from duty which would otherwise be collected if imported from any other country. **For details, obtain the leaflet *GSP & The Traveler* from you nearest Customs Office.** Many products of certain Caribbean countries are also exempt from duty under the Caribbean Basin Initiative (CBI). Most products of Israel may enter the U.S. either free of duty or at a reduced duty rate. Check with Customs. The U.S.-Canada Free Trade Agreement was implemented on January 1, 1989. U.S.

returning residents arriving directly or indirectly from Canada are eligible for free or reduced duty rates as applicable, on goods originating in Canada as defined in the Agreement.

Personal belongings of U.S. origin are entitled to entry free of duty. Personal belongings taken abroad, such as worn clothing, etc., may be sent home by mail before you return and receive free entry provided they have not been altered or repaired while abroad. These packages should be marked ***"American Goods Returned."*** When a claim of U.S. origin is made, this marking facilitates customs processing.

Foreign-made personal articles taken abroad are dutiable each time they are brought into our country unless you have acceptable proof of prior possession. Documents which fully describe the article, such as a bill of sale, insurance policy, jewelers appraisal, or receipt for purchase, may be considered reasonable proof of prior possession. Items, such as watches, cameras, tape recorders, or other articles which may be readily identified by serial number or permanently affixed markings, maybe taken to the Customs office nearest you and registered before your departure. **The Certificate of Registration provided will expedite free entry of these items when you return.** Keep the certificate as it is valid for any future trips as long as the information on it remains legible. Registration cannot be accomplished by phone nor can blank registration forms be given or mailed to you to be filled out at a later time.

Payment of Duty, required at the time of your arrival on articles accompanying you, may be made by any of the following ways: 1) U.S. currency (foreign currency is not acceptable); 2) Personal check in the exact amount of duty, drawn on a national or state bank or trust company of the U.S., made payable to the "U.S. Customs Service"; 3) Government check, money order or traveler's checks are acceptable if they do not exceed the amount of the duty by more than $50. Second endorsements are not acceptable. ID must be presented, e.g., traveler's passport or driver's license; and 4) In some locations you may pay duty with credit cards from Discover, MasterCard or Visa.

A complete booklet of customs hints for returning U.S. citizens, *"Know Before You Go,"* is free by writing the Department of the Treasury, U.S. Customs Service, Washington, DC 20229. Ask for Customs Publication No. 512.

Also be aware that certain souvenirs commonly available abroad may not be legally imported into the U.S. Several U.S. laws and an international treaty, designed to combat excessive exploitation of endangered species, make it a crime to bring many wildlife souvenirs back to the U.S. Be alert for certain reptile skins and leathers, depending on their country of origin, live birds and bird feathers; ivory from Asian and African elephants; certain plants, and fur from spotted cats, marine mammals, polar bears. There are many others.

For more information and a pamphlet called *Buyer Beware!*, write the Division of Law Enforcement, U.S. Fish and Wildlife Service, P.O. Box 3247, Arlington, VA 22203-3247.

Many food products are restricted or prohibited from entry into the U.S.; food items can harbor foreign pests and diseases that can damage American crops and livestock. Be sure of the restrictions before attempting to carry any food or meat products into the U.S. Check you local phone book for the nearest office of the U.S. Department of Agriculture, Animal and Plant Health Inspection Service.

If you plan to buy trademarked items while abroad, be aware of trademark restrictions on certain manufactured products. Although many trademark owners do not place restrictions on the number of goods a traveler may import, it is an owner's right to do so, even if goods are for the traveler's own personal use. Some commonly purchased trademarked items are: cameras and other optical goods, audio and video equipment, jewelry and precious metalware, perfume and like products. ***Trademark Information for Travelers* (Customs Publication No. 508) provides the most recent list of potential trademark restricted items.** Write to the Department of the Treasury, U.S. Customs Service, Washington, DC 20229.

SPECIAL CUSTOMS HIGHLIGHTS FOR CIVILIAN AND MILITARY GOVERNMENT PERSONNEL

The U.S. Customs Service is responsible for clearing all merchandise entering the United States. All imported goods are subject to a customs duty unless specifically exempted from this duty by law. Persons arriving in the U.S. from foreign countries are classified for customs purposes as either residents of the United States or nonresidents. Certain exemptions from payment of duty on the articles brought with them are provided.

A special provision allows U.S. Government personnel (military and civilian) to enter their personal and household effects without payment of tax when returning from an extended duty assignment overseas. Should they return to the U.S. for purposes of leave or TDY before their overseas assignment is concluded, they may claim the customs status of either a returning resident or a nonresident. Members of their family residing with them also may claim either status when returning for a short visit. The classification and rates of duty, or exemptions therefrom, on imported goods are governed by the Tariff Schedules of the United States (TSUS). Under item 817.00 of the Tariff Schedules, personal and household effects of any person (military or civilian) employed by the U.S. Government, and members of his family residing with him at his post or station, may be entered free of duty unless items are restricted, prohibited, or limited as in the case of liquor and tobacco.

To claim this exemption, the person in the service of the United States must be returning to the States under Government order upon termination of an assignment to extended duty outside the Customs territory of the U.S.

An assignment to extended duty abroad must be of at least 140 days duration, except as noted for Navy personnel. Military and civilian personnel are entitled to free entry privileges, if: 1) They are returning, at any time, upon termination of an assignment of extended duty; or **2**) They are under permanent change of station orders to another post or station abroad, necessitating return of their personal and household effects to the United States.

Navy personnel serving aboard a United States naval vessel or supporting a naval vessel when it departs from the U.S. on an intended deployment of 120 days or more outside the country and who continue to serve on the vessel until it returns to the U.S. are entitled to the extended duty exemption.

Free entry of accompanied and unaccompanied effects of family members who have resided with the employee cannot be claimed under item 817.00 when imported before the employee's receipt of orders when terminating his extended duty assignment. Persons not entitled to this exemption: **1**) Employees of private businesses and commercial organizations working under contract for the U.S. Government; **2**) Persons under research fellowships granted by the United States Government; **3**) Peace Corps volunteers or employees of UNICEF; **4**) Persons going abroad under the Fulbright-Hayes Act of 1961 or under the Mutual Educational and Cultural Exchange Act of 1961. Item 817.00 applies, however, to any person evacuated to the United States under U.S. Government orders or instructions.

CUSTOMS DECLARATIONS

Accompanied Baggage: Articles which accompany you upon your return to the United States on PCS orders should be declared on Customs Form CF 6059B "Customs Declaration," if you travel on a commercial carrier. If you travel on a carrier owned or operated by the U.S. Government, including charter aircraft, you will execute **Department of Defense form DD 1854, "Customs Accompanied Baggage Declaration."** Be prepared to show the customs officer a copy of your travel orders.

Unaccompanied Baggage: If you are a DoD civilian or military member returning to the U.S. from extended duty overseas, you should execute **DD form 1252, "U.S. Customs Declaration for Personal Property Shipments,"** to facilitate the entry of your unaccompanied baggage and/or household goods into the U.S. A copy of your PCS orders, terminating your assignment to extended duty abroad, should accompany DD form 1252. This form is also used by a DoD sponsored or directed individual or employee of a nonappropriated fund agency which is an integral part of the military services.

All other Government employees should complete **"Declaration for Free Entry of Unaccompanied Articles," Customs form 3299, and attach a copy of their orders.**

By these declarations you certify that the shipment consists of personal and household effects which were in your direct personal possession while abroad and the articles are not imported for the account of another person or intended for sale. Employees completing CF 3299 must list restricted articles (e.g., trademarked items, firearms), and goods not subject to their exemption (e.g., excess liquor, articles carried for other persons) on the declaration and show the actual prices paid. DoD employees and military members whose shipment of personal and household effects are cleared by a military customs inspector (MCI) will indicate to the MCI any articles which are restricted or subject to customs duty. A notation will be made by the MCI on DD form 1252 and the shipment will be examined by U.S. Customs upon its arrival in the U.S. Shipments from other areas where MCIs are not assigned will be cleared upon arrival in the U.S.

Effects sent by mail are eligible for duty-free entry if the articles were in the returnee's possession prior to leaving the duty station. A copy of the government orders terminating the assignment must accompany the articles in a sealed envelope securely affixed to the outer wrapper of the parcel. **The parcel should also be marked clearly on the outside "Returned Personal Effects - Orders Enclosed."**

Articles taken with you from the United States need not be listed on your declaration. If such articles were repaired in a foreign country, list the cost of repairs. If the repaired or altered article is changed sufficiently to become a different article, it must be declared at its full value.

Effects sent home before your orders are issued, or purchased overseas and not delivered to you abroad but sent to your address in the States, do not qualify for free entry.

Merchandise of foreign origin purchased in overseas Post or Base Exchanges is subject to customs treatment and other import requirements and regulations, including trademark restrictions.

Merchandise of foreign origin purchased in overseas Post or Base Exchanges is subject to customs treatment and other import requirements and regulations, including trademark restrictions.

LIMITATIONS

Tobacco: Not more than 100 cigars may be imported free of duty as personal effects. There is no limitation on the numbers of cigarettes. Products of Cuban tobacco are prohibited to arriving U.S. citizens and residents, unless acquired in Cuba.

Liquor: Not more than four liters (135.2 fluid ounces) of alcoholic beverages, of which three liters must be bottles in the United States and of U.S. manufacture, may be imported free of duty as personal effects, if: **1)** It accompanies the employee or the member of his family making claim for entry at the time that person arrives in the U.S.; **2)** The member of the employee's family claiming the exemption is 21 years age or older (U.S. civilian or military personnel are exempt from the age requirement); **3)** The person requesting free entry does not claim the customs exemption for alcoholic beverages as a returning U.S. resident or nonresident.

No alcoholic beverages may be imported into the U.S. by mail nor can Customs release liquor in violation of the laws of the state where it is entered. As laws vary from state to state, this information may be obtained from state liquor authorities.

"Hide the jewelry in the blue suitcase Muriel, Hide the jewelry in the blue suitcase Muriel."

Drawing by artist Rockwell from Pacific Crossroads.

APPENDIX T: MILITARY AIRLIFT COMMAND (AMC) IN-FLIGHT FOOD SERVICE

No matter where you travel throughout the Air Mobility Command (AMC), you will find totally different, greatly improved flight meals. Flight menus have traditionally laced variety, quality and overall customer appeal. An exciting new in-flight food service program has been developed by the AMC Food Service Branch. The program was established to improve the quality, nutritional content, packaging, and presentation of flight meals within AMC.

Improvements to the food service program include the addition of "HealthyHeart" and breakfast menus, the exclusive use of deli meats, fresh fruits and vegetables, pasta salads, fruit cups, two-percent milk, cholesterol-free snacks, and whole wheat bread. "Junk food" items high in fats and sodium, such as candies and cream-filled pastries, are no longer served.

This food service and the in-flight kitchens that produce it are positioned throughout the world. All passengers ordering in-flight meals, even from Space-A terminals in exotic locales, will experience the same high-quality and nutritious food AMC now provides.

Following are some sample menus available on AMC flights. In-flight menu prices are established at the begining of each fiscal year (1 October) and may vary at different locations.

SANDWICH MEALS $2.50 - 3.00

When selecting you sandwich, please indicate the supplement packages you would like to compliment your meal. Diet soft drinks may be substituted upon request. Menus subject to change due to nonavailability.

SANDWICH MENUS

1. Turkey and American Cheese Hoagie
2. Turkey, Ham & Swiss Cheese Hoagie
3. Ham, Roast Beef Hoagie
4. Ham, American Cheese Hoagie
5. Roast Beef, Swiss on Whole Wheat
6. Ham, Corned Beef & American Cheese on Whole Wheat
7. Ham & Swiss Cheese on Rye
8. Turkey, ham & American Cheese on Whole Wheat
9. Ham & Provolone Cheese on Whole Wheat10. Corned Beef and Swiss Cheese on Rye

*11. Turkey on Whole Wheat at
*12. Peanut Butter & Jelly on Whole Wheat
13. Ham, Roast Beef, and American Cheese Kaiser
14. Turkey and Swiss Cheese on Whole Wheat
15. Fried Chicken with Dinner Roll
* 16 Baked Chicken with Dinner Roll

*HEALTHY HEART MENUS

STANDARD SUPPLEMENT PACKAGE

A. Fruit Cup
Vegetable Tray
Assorted Veg Bread
Assorted Snack Item
Soft Drink and Juice
Condiments, Flight Pack

B. Fresh Fruit
Italian Veg Pasta Salad
Raisins
Snack Pack Pudding
Lowfat White Milk & Juice
Condiments, Flight Pack

C. Fresh Fruit
Vegetable Tray
Snack Pack Pudding
Raisins
Soft Drink and Juice
Condiments, Flight Pack

D. Fruit Cup
Italian Veg Pasta Salad
Assorted Snack Item
Assorted Veg Bread
Lowfat White Milk & Juice
Condiments, Flight Pack

SNACK MEALS $1 - 1.50

SNACK MENUS
A. Fried or Baked Chicken
Vegetable Tray
Fresh Fruit
Dinner Roll w/Margarine
Lowfat Milk
Condiments, Flight Pack

B. Ham-n-Cheese on Wheat
Fresh Fruit
Vegetable Tray
Danish
Lowfat Milk
Condiments, Flight Pack

C. Turkey, Ham and Swiss Cheese on Whole Wheat
Fresh Fruit
Vegetable Tray
Soft Drink
Condiments, Flight Pack

D. Comed Beef, Swiss on Rye
Italian Veg Pasta Salad
Fresh Milk
Lowfat Milk
Condiments, Flight Pack

E. Roast Beef and Provolone Hoagie
Italian Veg Pasta Salad
Fresh Fruit
Fruit Juice
Condiments, Flight Pack

HEALTHY HEART MENUS
F. Chef Salad
Diet Dressing Crackers
Fresh Fruit
Skim Milk
Flight Pack

G. Baked Chicken
Fresh Fruit
Vegetable Tray
Raisins
Vegetable Juice
Flight Pack

H. Turkey and Swiss Cheese on Whole Wheat
Vegetable Tray
Fresh Fruit Skim Milk
Condiments, Flight Pack
Fresh Fruit Skim Milk
Condiments, Flight Pack

I. Tuna (unprepared)
Salad Dressing
Wheat Bread
Vegetable Tray
Fresh Fruit
Vegetable Juice
Flight Pack

BREAKFAST MENU $1 - 1.50

J. Breakfast Cereal
Fresh Fruit
Danish
Lowfat Milk, Fruit Juice
Yogurt
Flight Pack

K. Ham and Swiss Cheese on Bagel
Fresh Fruit Danish
Fruit Juice
Condiments, Flight Pack

New lockers at Rhein-Main AB, Germany

Morale, Welfare, and Recreation (MWR) operates these coin-operated lockers.

SPACE-A REMOTE SIGN-UP PULLOUT SHEET

APPLICATION FOR SPACE AVAILABLE MILITARY AIR TRAVEL
(VIA TELEFAX (FAX) OR UNITED STATES MAIL (USPS)

TO:________________________
Passenger Terminal/Station

____
Unit/Street Address

____
Base/Installation Name

____
Country/City, ZIP/APO, FPO

____
Passenger (Sponsor) Name
(last, first, middle initial)

____
I. D. Card No. & Expir. date

FAX:________________________
Telefax No. in CONUS use
Area Code + NO., in OCONUS
use Country/City Code + NO.

____
Pay Grade/Title Total Seats Req

____
Active Duty, Res Comp, Retired (check)

____
Passport No.(If Req) & Expir. date

ACCOMPANYING DEPENDENTS

____
(1) Dependents Name and
Relationship to Sponsor

____
I. D. Card No. & Expir. date

____
Passport No. & Expir. date

____
(2) Dependents Name and
Relationship to Sponsor

____
I. D. Card No. & Expir. date

____
Passport No. & Expir. date

____
(3) Dependents Name and
Relationship to Sponsor

____
I. D. Card No. & Expir. date

____
Passport No. & Expir. date

____
(4) Dependents Name and
Relationship to Sponsor
(Use 2nd Sheet for more Dependents)

DESTINATION SELECTION

List below a maximum of five (5) Country or OCONUS destinations, in your order of preference, for air travel. You may use the fifth destination as "all" to take advantage of opportune airlift:

(1)______________(2)______________

(3)______________(4)______________

(5)______________

Passengers traveling in CONUS only, list CONUS as only destination.

OTHER DOCUMENTATION

Active Duty: Must telefax (FAX) with this Application for Air Travel, a signed/authenticated copy of their Service Leave/Pass Form (AF Form 988, DA Form 31, NAVCOMP 3065, & NAVMC-3 and others from USCG, USPHS & NOAA).

Reserve Components: Must telefax (FAX) with this Application for Air Travel, a current (not more than 180 days old) copy of your signed/ Authenticated, DD Form 1853, "Authentication of Reserve Status for Travel Eligibility".

Note: All I. D. cards and passports must be signed.

ABBREVIATIONS

Comp=Components
CONUS=Continental United States
Expir=Expiration
NOAA=National Oceanic & Atmospheric Administration
OCONUS=Outside Continental United States
Res=Reserve
USCG=United States Coast Guard
USPHS=United States Public Health Service

BORDER CLEARANCE

I certify that the above listed border clearance document (and others documents as required) for myself and my dependents are current and meet the DoD Foreign Clearance Guides, Personnel Entrance Requirements of the countries to which I have applied for Air Travel. (Consult air departure terminals and personnel offices issuing orders for information).

____

Signature date

____

Print Name, Grade/Title, Service

Note: The actual sign-up date (application for Air Travel) will be the transmission date and time (converted to Julian Date) of the telefax (FAX) transmission to the terminals; the date and time of receipt of mail applications to the terminals or; applications in person at the terminals.

All applicants will be placed on the Space Available rooster in the order of application within in each priority and category of travel. They will remain on the list for 45 days or the duration of their approved leave which ever expires first.

This Remote Sign-up initiative has been documented in a change to Chapter 4, Space Available Passengers, DoD 4515.13-R.

This Remote Sign-Up form was prepared for subscribers to Military Living's R & R Space-A Report®, an all ranks military travel newsletter (Military Living Publications, P. O. Box 2347, Falls Church, VA 22042-0347, Tel: 703-237-0203). Military Living assumes NO responsibility as to the absolute correctness or suffiency of any representation contained in this blank or completed form. The final authority for your air travel remains with the Service/Military air passenger terminal from which you plan to depart. Please call your chosen terminal to insure that you have complied with their requirements.

Roy Crawford, Sr. tries out the famous C5A ladder which leads to the passenger compartment. See picture of a "stair truck" on the next page. Stair trucks are not always available.

— Photo by Ann Crawford

Stair trucks are a little easier to climb but are not always available.

— U.S. Army Photo by Larry Lane, Soldiers Magazine

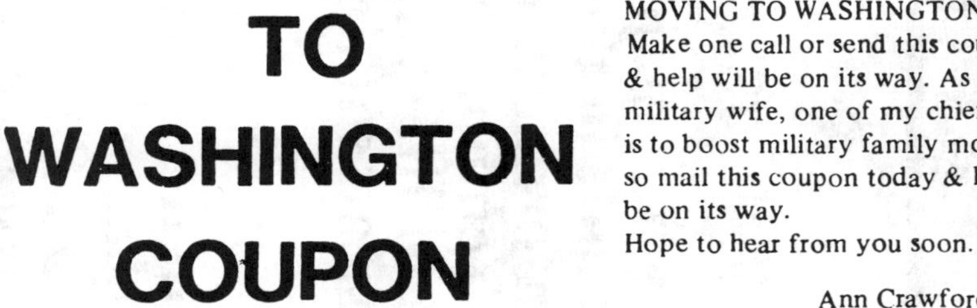

MOVING TO WASHINGTON COUPON

"One Shot" Help

MOVING TO WASHINGTON?
Make one call or send this coupon & help will be on its way. As a military wife, one of my chief goals is to boost military family morale . . . so mail this coupon today & help will be on its way.
Hope to hear from you soon.

Ann Crawford, Publisher
Military Living Magazine

TO: Mrs Ann Crawford, Publisher, Military Living
P.O. Box 2347, Falls Church, VA 22042-0347
Phone: (703) 237-0203, FAX (703) 237-2233

Our Family Is:
- ☐ Army
- ☐ Navy
- ☐ Air Force
- ☐ Marine
- ☐ Coast Guard
- ☐ P.H.S.
- ☐ NOAA
- ☐ Other
- ☐ Active
- ☐ Retired
- ☐ 100% DAV

Military member's name/rank ______________________________

______________________ Spouses Name ______________________

Address: ______________________________

City/State/Zip: ______________________________

Tel: ______________ Total number in family: ______
Number of children: ______

Area assigned to: ______________________________

We expect to arrive: ______________________________

We would like info, if possible, on the following:
- ☐ House ☐ Apt. ☐ Condo
- ☐ Renting ☐ Buying
- ☐ Car
- ☐ Furniture
- ☐ Major Appliances

Short Term Housing
- ☐ Military Lodging
- ☐ Hotel/Motel ☐ B & B
- ☐ Short Term Apt.

Type of job: ______________________________

____

- ☐ Banking/Checking Accounts
- ☐ Employment Opportunities for spouse
- ☐ Real Estate Career Opportunities
- ☐ Legal Services/Settlement Atty.
- ☐ College Opportunities
- ☐ Investment Opportunities
- ☐ Travel in nearby areas
- ☐ Back Issues of **Military Living**
- ☐ Window Treatments
- ☐ Dentist ☐ Doctor

AT TIMES LIKE THIS........

At times like this, when major changes are being made in the Uniformed Services, information on Space-A air travel, Temporary Military Lodging and Military RV, Camping & Rec Areas can suddenly change.

If you want to be in the know before most everyone else, subscribe to *Military Living's R&R Space-A Report* ®! Breaking news will be carried in this six time yearly all ranks travel newsletter which is available by subscription only. To subscribe, see the coupons in this book or call (703) 237-0203 for information or to order with VISA, MasterCard or American Express.

CENTRAL ORDER COUPON

Military Living Publications
P.O. Box 2347, Falls Church VA 22042-0347
TEL: (703) 237-0203 FAX: (703) 237-2233

Publications	QTY
R&R Space-A Report®. *The worldwide travel newsletter.* 6 issues per year. 5 yrs/$47.00 - 2 yrs/$22.00 - 3 yrs/$30.00 - 1 yr/$14.00	
Military Space-A Air Basic Training. $11.00	
Military Space-A Air Opportunities Air Route Map. (Folded) $10.00	
Military Space-A Air Opportunities Around the World. $15.95	
Temporary Military Lodging Around the World. $13.95	
Military RV, Camping & Rec Areas Around the World. $11.95	
U.S. Forces Travel and Transfer Guide, USA and Caribbean Areas. $11.95	
U.S. Military Museums, Historic Sites & Exhibits. (Soft Cover) $16.95 - (Hard Cover) $26.95	
United States Military Road Atlas $16.95	
U.S.Military Installation Road Map. (Folded) $6.50	
United States Military Medical Facilities Map (Folded) $6.95	
COLLECTOR'S ITEM! Desert Shield Commemorative Maps. (Folded) $7.00 (2 unfolded wall maps in a hard tube) $16.00	
Assignment Washington: Military Road Atlas. Maps & Charts of Washington Area Military Installations. $8.95	
California State Military Road Map ALL NEW 1994! - (Folded) $4.95 **Florida State Military Road Map** ALL NEW 1994! - (Folded) $4.95 **Mid-Atlantic States Military Road Map** ALL NEW 1994! - (Folded) $4.95 **Texas State Military Road Map** ALL NEW 1994! - (Folded) $4.95	
Military Living Magazine, Camaraderie Washington. *Local Area magazine.* 1 year (12 issues) 1st Class Mailing $10.00	
Virginia Addresses add 4.5% sales tax (Books, Maps, & Atlases only) **TOTAL $**	

*If you are an R&R Space-A Report® subscriber, you may deduct $1.00 per book. (No discount on the R&R Report itself or on the maps or atlases.) For 1st Class Mail, add $1.00 per book or map. Mail Order Prices are for U.S. APO & FPO addresses. Please consult publisher for International Mail Price. Sorry, no billing. GREAT FUND RAISERS! Please write for wholesale rates.
We're as close as your telephone...by using our Telephone Ordering Service. We honor VISA, MasterCard, and American Express. Call us at **703-237-0203 (Voice Mail after hours)** or Fax 703-237-2233 and order today! Sorry, no collect calls. Or...fill out and mail the order coupon below. Thank You!

NAME:__________
STREET:__________
CITY/STATE/ZIP:__________
PHONE:__________ SIGNATURE:__________
RANK (or rank of sponsor):__________ Branch Of Service:__________
Active Duty:____ Retired:____ Widow/er:____ 100% Disabled Veteran:____ Guard:____ Reservist:____ Other:____
Card #__________ Card Expiration Date:__________

Mail check/money order to Military Living Publications, P.O. Box 2347, Falls Church, VA 22042-0347 - **Tel: 703-237-0203** - FAX: 703-237-2233.
Prices subject to change. Please check here if we may ship and bill difference.

The Space-A desk is a popular place! — *Photo by SSgt Larry'Lane, Soliders Magazine*

CENTRAL ORDER COUPON
Military Living Publications
P.O. Box 2347, Falls Church VA 22042-0347
TEL: (703) 237-0203 FAX: (703) 237-2233

Publications	QTY
R&R Space-A Report®. *The worldwide travel newsletter.* 6 issues per year. 5 yrs/$47.00 - 2 yrs/$22.00 - 3 yrs/$30.00 - 1 yr/$14.00	
Military Space-A Air Basic Training. $11.00	—
Military Space-A Air Opportunities Air Route Map. (Folded) $10.00	
Military Space-A Air Opportunities Around the World. $15.95	
Temporary Military Lodging Around the World. $13.95	
Military RV, Camping & Rec Areas Around the World. $11.95	
U.S. Forces Travel and Transfer Guide, USA and Caribbean Areas. $11.95	
U.S. Military Museums, Historic Sites & Exhibits. (Soft Cover) $16.95 - (Hard Cover) $26.95	
United States Military Road Atlas $16.95	
U.S.Military Installation Road Map. (Folded) $6.50	
United States Military Medical Facilities Map (Folded) $6.95	
COLLECTOR'S ITEM! Desert Shield Commemorative Maps. (Folded) $7.00 (2 unfolded wall maps in a hard tube) $16.00	
Assignment Washington: Military Road Atlas. Maps & Charts of Washington Area Military Installations. $8.95	
California State Military Road Map ALL NEW 1994! - (Folded) $4.95 **Florida State Military Road Map** ALL NEW 1994! - (Folded) $4.95 **Mid-Atlantic States Military Road Map** ALL NEW 1994! - (Folded) $4.95 **Texas State Military Road Map** ALL NEW 1994! - (Folded) $4.95	
Military Living Magazine, Camaraderie Washington. *Local Area magazine.* 1 year (12 issues) 1st Class Mailing $10.00	
Virginia Addresses add 4.5% sales tax (Books, Maps, & Atlases only) **TOTAL $**	

*If you are an R&R Space-A Report® subscriber, you may deduct $1.00 per book. (No discount on the R&R Report itself or on the maps or atlases.) For 1st Class Mail, add $1.00 per book or map. Mail Order Prices are for U.S. APO & FPO addresses. Please consult publisher for International Mail Price. Sorry, no billing. GREAT FUND RAISERS! Please write for wholesale rates.
We're as close as your telephone...by using our Telephone Ordering Service. We honor VISA, MasterCard, and American Express. Call us at **703-237-0203 (Voice Mail after hours)** or Fax 703-237-2233 and order today! Sorry, no collect calls. Or...fill out and mail the order coupon below. Thank You!

NAME:__________
STREET:__________
CITY/STATE/ZIP:__________
PHONE:__________ SIGNATURE:__________
RANK (or rank of sponsor):__________ Branch Of Service:__________
Active Duty:____ Retired:____ Widow/er:____ 100% Disabled Veteran:____ Guard:____ Reservist:____ Other:____
Card #__________ Card Expiration Date:__________

Mail check/money order to Military Living Publications, P.O. Box 2347, Falls Church, VA 22042-0347 - **Tel: 703-237-0203** - FAX: 703-237-2233.
Prices subject to change. Please check here if we may ship and bill difference.

Ann Crawford talks with a Senior Master Sergeant at Ramstein's old air passenger terminal in 1993. The new terminal is now open.